PRAISE

Fire on the

T0115992

"*Fire on the Mountain* weaves together a tense narrative of almost cinematic action, starring ballsy cowboy smoke jumpers. . . . Maclean's well-sketched personalities bring the action on the ground convincingly to life—and knowing up front that many of his main characters won't survive South Canyon makes this tragic tale that much more compelling."   —Amazon.com

"A finely wrought factual and emotive record."   —*Booklist*

"The mix of politics and firefighting, the interaction of hot emotions and cool technical minutiae, the ominous intermingling of past and present—all make for a gripping scene, one of dozens in [this] excellent book. . . . The most complete accounting yet of what happened both on the mountain and behind the scenes."
                                    —*Rocky Mountain News*

"*Fire on the Mountain* paints a vivid portrait. . . . Maclean's work is not just investigative journalism. It is a memorial."
                                    —*High Country News*

"Maclean's carefully crafted exposé honors those dead. . . . He has a reporter's tested punch."   — *Kirkus Reviews*

"[Maclean is] a fine storyteller."   —*Washington Post Book Review*

"Be forewarned: Once you start reading this book it is very hard to put down. [Maclean] is able to provide the kind of information that people have wanted for the last five years, the kind of information that can save other lives. *Fire on the Mountain* is a must-read."
                                    —*Wildland Firefighter* magazine

Dan Jackson

ABOUT THE AUTHOR

JOHN N. MACLEAN's *Fire on the Mountain* was a Mountains and Plains Independent Booksellers Association's best nonfiction title of 2000. A newspaper reporter and longtime student of wildfire, he is the author of *The Thirtymile Fire* and *Fire and Ashes*, a *Chicago Tribune* "Best Book" of 2003. He also assisted in the posthumous publication of *Young Men and Fire,* a work of nonfiction by his father, Norman Maclean. He divides his time between Washington, D.C., and Montana.

FIRE ON THE MOUNTAIN

FIRE ON THE MOUNTAIN

*The True Story
of the South Canyon Fire*

JOHN N. MACLEAN

HARPER ● PERENNIAL

NEW YORK ● LONDON ● TORONTO ● SYDNEY ● NEW DELHI ● AUCKLAND

HARPER ● PERENNIAL

First Harper Perennial edition published 2009.

The Library of Congress has catalogued the hardcover edition as follows:
Maclean, John N.
    Fire on the mountain : the true story of the South Canyon fire / John N. Maclean
        p. cm.
Originally published : 1st ed. New York : William Morrow, 1999.
ISBN 0-7434-1038-6
1. Wildfires—Colorado—Garfield County—Prevention and Control.
I. Title.
SD421.32.C6M235   2000
363.37'9—dc21
                                                        00-036317

ISBN 978-0-06-182961-1 (pbk.)

24 25 26 27 28 LBC 18 17 16 15 14

To my wife, Frances, with love

Fire on the mountain,
Run boys run;
Fire on the mountain,
Run girls run.
Take your partner as you go,
Jim along Josie,
Jim along Joe.

—From "Fire on the Mountain," an
early American play-party song

Storm King Mountain

4:00 P.M., July 6, 1994

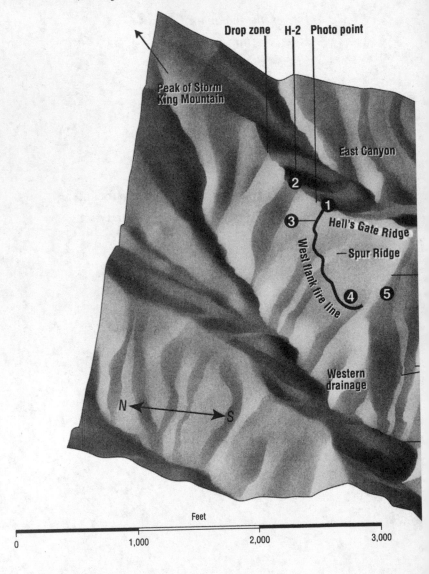

Drop zone H-2 Photo point

Peak of Storm King Mountain

East Canyon

2

1

3

Hell's Gate Ridge

Spur Ridge

West flank fire line

4   5

Western drainage

N ←→ S

Feet

0   1,000   2,000   3,000

1 The incident commander, Butch Blanco; Michelle Ryerson, and her crew — Todd Abbott, Jim Byers, Eric Christianson, Mike Hayes, Loren Paulson, Brian Rush, and Neal Shunk; and ten Prineville Hot Shots — Bryan Scholz, foreman, and Bill Baker, Kip Gray, Brian Lee, Tony Johnson, Louie Navarro, Tom Rambo, Alex Robertson, Mike Simmons, Kim Valentine.

2 The helitacks, Rich Tyler and Rob Browning, and the Prineville superintendent, Tom Shepard.

3 The tree where these firefighters met: Sunny Archuleta, Derek Brixey, Sarah Doehring, Kevin Erickson, and Brad Haugh.

4 The smoke jumper in charge, Don Mackey; three other jumpers — Eric Hipke, Roger Roth, and Jim Thrash — and nine Prineville Hot Shots — Kathi Beck, Tami Bickett, Scott Blecha, Levi Brinkley, Doug Dunbar, Terri Hagen, Bonnie Holtby, Rob Johnson, and Jon Kelso.

5 Dale Longanecker, Tony Petrilli, and seven other smoke jumpers — Mike Cooper, Mike Feliciano, Quentin Rhoades, Sonny Soto, Eric Shelton, Billy Thomas, and Keith Woods.

H-1

Escape route

Lunch
Spot Ridge

Point
of origin

Double Draws

I-70

Origin
of blowup

Denver and
Rio Grande
Western railway

Colorado River

© By Joe Lertola for John N. Maclean

Part

I

ONE

THE CITY OF Grand Junction, located at the confluence of the Colorado and Gunnison rivers, is the crossroads of western Colorado for trade, agriculture and government. The fertile river valley is ringed by flat-topped mesas, lonely, bleak and arid in the best of times. In the summer of 1994 the city's 38,000 residents and all of western Colorado, from the mountain resorts to the peach orchards around Grand Junction, faced a severe drought. There had been little snowfall over the winter, and unusually hot, dry weather had followed in May and June. Only once every thirty or forty years did such events occur in combination—drier conditions in 1990 had not lasted as long.

In anticipation of an outbreak of forest fires Christopher J. Cuoco, a National Weather Service forecaster from Denver, had been stationed in Grand Junction at the Western Slope Coordination Center, a federal fire office located at the city's Walker Field airport. At 3:00 P.M. on Friday, July 1, 1994, Cuoco issued a Red Flag Watch to firefighters in western Colorado. Expect high winds, lightning and no rain, Cuoco told them, and added, "A high potential for large fire growth."

By six-thirty the next morning, Saturday, July 2, Cuoco had upgraded the watch to a Red Flag Warning, meaning that the chance of a major storm had become, for him, a certainty. In his new forecast, Cuoco told firefighters, "Stronger winds will come later this afternoon. Widely scattered thunderstorms with little or no rain. Wind gusts to forty miles an hour possible."

Cuoco's prediction joined a stream of drought-related warnings issued by the array of government agencies charged with forest-fire suppression, from the U.S. Forest Service and Bureau of Land Management (BLM) to state, county and volunteer fire offices. The cautions ran a broad gamut: One federal safety officer reminded fire crews to drink plenty of water "and not so much sugar drinks"; another warned that a wind of only ten miles an hour could play havoc with fire in the thickets of scrub oak, pinyon pine and juniper in the high desert country around Grand Junction.

All these levels of government—federal, state and county—had anticipated trouble early in the spring when several fires deliberately set to clear brush had burned out of control. After a series of meetings, the Bureau of Land Management, responsible for more acreage than any other agency, announced an aggressive policy of attacking all fires as soon as they were spotted. By the end of June, with precipitation at half normal levels and temperatures rising over the hundred-degree mark even in the mountains, virtually every forestry worker in Colorado knew that an electrical storm like the one Cuoco forecast could literally spark disaster.

The first bolt of lightning struck on July 2 in late morning, about 11:40. Before the day ended, more than fifty-six hundred strikes would follow, making this one of the worst lightning storms in the history of the state. Whatever rain it carried evaporated long before reaching ground. The storm arrived at Grand Junction at midafternoon, rattling metal roofs, sending trash bins end over end and scooping bone-dry dirt and sand into an old fashioned, Depression-era dust storm.

At Walker Field airport, the wind sent stacks of tumbleweed hurtling along runways. The sky turned an eerie gray-brown. Cuoco rushed outside and began snapping photographs.

The thunderstorm swept through the Grand Valley of the Colorado River, fifty miles long and twenty miles wide. It continued east to the Book Cliffs, a line of mesas named for their resemblance to a shelf of books, and poured through a gap cut by the Colorado River, one of the nation's longest waterways and one of the most heavily used—for everything from electrical-power generation to float trips. It followed upriver a dozen miles on a twisting course through Debeque Canyon, barely wide enough to hold the river and the four lanes of Interstate 70, the fast new highway linking Denver and the western part of the state. Centuries of winds had carved the canyon's sandstone cliffs into a pattern of small, deep hollows called "honeycomb weathering," easily mistaken for swallows' nests.

The storm rumbled out of Debeque Canyon and onto a broad plain stretching eastward more than thirty miles to the beginning of the Rocky Mountains. Once again it kept on course with the river, following the Colorado to Battlement Mesa, a massive formation standing alone in the valley twenty miles beyond Debeque Canyon. There, in July 1976, three men had been killed battling a forest fire, the last such deaths in the state. Fresh growth had long since covered the burn scars, but for those who knew its history Battlement Mesa bore continuing witness to what can happen on a hot afternoon on a steep slope in the wake of a lightning storm.

From the mesa the storm had an open path to the Grand Hogback, a bristling hundred-mile-long ridge that runs at right angles to the Colorado River and marks the official western boundary of the Rocky Mountains. The storm cascaded over the low ridge, hardly breaking stride, and entered the final broadening of the valley. Ahead the mountains rose steeply, but the vegetation remained the scrubby Gambel oak and pinyon pine and juniper, or PJ, of the desert. The snowcapped peaks and alpine forests of picture-postcard Colorado lay farther east.

A dozen miles beyond the Grand Hogback, nearly ninety miles in all from Grand Junction, the mountains came together in what appeared to be an unbroken front, though on closer inspection a narrow V could be seen barely wide enough for the Colorado River, I-70 and a set of railroad tracks. To the north of the Col-

orado rose a single bulky peak almost nine thousand feet high—
Storm King Mountain.

The mountain had a remote, unapproachable appearance despite overlooking a major highway. Ridges fanned out from its
peak like arms from a misshapen octopus. One ridge, the longest
and heaviest, had been sheared off by the river and now loomed
above the highway. It came to be called by different names; in
this story it will be Hell's Gate Ridge. The slopes of the mountain
and its ridges were covered with scrub oak, shining an oily green,
broken by stands of dwarfish PJ. Rockslides exposed dull tan and
reddish shale. Dead brush filled deeply eroded gullies.

When the thunderstorm reached the end of the valley, it hurled
itself against Storm King Mountain and caught there.

David and Jo Temple were at dinner with their two sons—
Matthew, ten, and Beau, seven—when the storm hit. The Temple house, built in the foothills of the mountain, resembled an
old-fashioned fire-lookout tower, a role it came to play in reality
as events unfolded. It was two stories high, made with rough
wooden siding, and had a walkway around the glassed-in second
story. All it needed to complete the picture was a forest ranger
wearing a Smokey Bear hat and sporting a pair of binoculars.

The Temple family had come out from the Midwest two decades earlier in search of a place of their own in the mountains.
Their house had a sweeping view across a meadow to the heights
of Storm King. In their first years there, the Temples watched on
winter evenings as herds of elk fed in the meadow, the animals
tolerant of the whine from an occasional vehicle on the two-lane
highway—I-70 before it became a four-lane interstate. But as
time passed, the Temples' dream vanished, piece by piece.

In the 1980s developers acquired the meadow, putting up a
ring of custom-built houses called Canyon Creek Estates. With a
nod to nature, they left enough of the meadow to form an oval
common in the center of the development, but the elk no longer
gathered there to browse. The last link of I-70 was completed in
1992, bringing new life for a population boom already under way.

What had been called a "home in the woods" became known
instead as the "urban-wildland interface," and since 1985, more
than three hundred such structures had been lost to fire every

year across the West. A fire had burned to the Temples' rear doorstep in 1990 after igniting on a hill behind their house. Fire engines pulled up within minutes, but their crews had no chance against a blaze with the wind at its back; the Temples were about to lose everything. With flames less than a hundred yards from the house, the wind shifted and the fire raced off on a different course. Charred tree trunks marking its route can still be seen.

Now, in the midst of the 1994 storm, Jo Temple tried to calm her children. Outside, storm and mountain had joined together, an indistinguishable mass of gray and black fractured by bolts of lightning. A single bolt made the link between sky and earth, striking a tree on Storm King's heaviest ridge, Hell's Gate Ridge, at the tip overlooking the Colorado River that came to be called Hell's Gate Point. The bolt tore open the tree and plunged into heartwood.

By the time the storm had spent itself and rolled off Storm King Mountain, it had ignited fifteen fires in the Grand Junction District of the Bureau of Land Management, which includes Storm King and a surrounding area of 1.8 million acres, a heavy burden for a district that normally sees fewer than eighty fires a season.

By Sunday morning, July 3, the only evidence on the mountain of the storm's furious embrace was a puff of smoke on Hell's Gate Point above I-70. In days to come, homeowners watched it from their backyards in Canyon Creek Estates, one making a video that not only showed the fire's progress but carried the homeowner's narration, in which he uncannily predicted the outcome of the drama days before it happened. Motorists by the score pulled off I-70 to take photographs from the fire's first days as a smoke signal to its eventual, infernal climax; boaters watched it from the Colorado River, railroaders from the tracks across the river.

In the spa town of Glenwood Springs, tucked alongside the river five and a half miles to the east as the crow flies, bathers in the Great Swimming Pool, fed by hot springs legendary for their curative powers, saw the rising smoke come "over the King," just as residents for more than a century had watched storms come across the mountain's heights.

The smoke should have set sirens wailing from Grand Junction, with its many federal administrative offices, across the country— to the suburbs of Denver, where the Rocky Mountain Coordination Center, part of a national system for fighting fire, had its headquarters, and to Boise, Idaho, home of the parent National Interagency Fire Center.

If the smoke column wasn't enough, a daily situation report by the BLM Grand Junction District should have prompted someone in the federal complex to action. On July 3 the district cautioned:

"Red flag warning, no relief in sight.

"Prognosis: Local I.A. [initial attack] is spread thin, radio communication inadequate for fire load and safety is in jeopardy."

Not that the situation was ignored—quite the contrary. Over the next two days, a parade of smoke jumpers, fire-engine crews, volunteer firemen, fire managers and aerial observers made almost hourly excursions to Storm King Mountain. They arrived in airplanes, fire engines, police squad cars and sport utility vehicles. They located the fire's exact position on maps, debated who should fight it, measured its rate of spread, assessed its future and ranked it against all others in the district.

But when dawn rose two days later on Tuesday, July 5, with the fire grown to a defiant threat, no one—not a single soul— had walked to the fire, let alone fought it.

It was still dark the morning after the storm, Sunday, July 3, when Chris Cuoco arrived at the Western Slope Coordination Center located in a simple metal-sided building at Grand Junction's Walker Field. At the center a handful of people were responsible for air support for fires in western Colorado and a slice of eastern Utah. Alone at this hour, except for an overnight radio dispatcher, Cuoco began to gather up limp facsimile paper containing weather data, but halted at a map showing the lightning pattern from the previous evening. The strikes had been so dense that a solid black line overlay the Colorado River. "In seven years in Colorado I'd never seen anything like it; it was one of the most energetic storms I'd *ever* seen," Cuoco said later. "That was one helluva storm."

At the age of thirty-seven, Cuoco retained a sense of wonder about the sky, coupled with a scientist's objectivity. He had the look of a cheerful monk—plump, open-faced and radiating goodness, but with a sharp eye. His first job as a forecaster had been as an Air Force lieutenant based inside Cheyenne Mountain, the nuclear-bomb-proof hideaway near Colorado Springs, assigned to the North American Aerospace Defense Command, or NORAD. There, he had monitored weather conditions for NORAD bases around the globe.

On one occasion he had found himself at the end of a command chain that linked him to a one-star general, his boss, who was talking to a four-star general, the chief of the Strategic Air Command, who was talking to the National Security Adviser, who was reporting to the President of the United States. The question came down to Cuoco: Are the streamers we see over the North Atlantic Ocean caused by the aurora borealis, or should we consider intercontinental ballistic missiles? Cuoco's answer was relayed back to the president: Yes, sir, they are the aurora borealis.

After leaving the Air Force, Cuoco spent two years in a Roman Catholic seminary, followed by a year of church work with youth, and then he took a job with the National Weather Service, serving his internship in Grand Junction. By 1994 he had transferred to Denver and become head of the service's fire weather program for the state of Colorado.

Cuoco was due to produce the day's first forecast by 6:00 A.M., still an hour and a half away. He had taken up duties at the Western Slope center a few days earlier, a last-minute idea of the center manager, Paul Hefner, who wanted a weather expert at his elbow. But it was an experiment. Cuoco had quickly discovered that no reliable procedure existed to pass along his forecasts to firefighters. The forecasts were as likely to wind up thumbtacked to a wall as to be transmitted by radio dispatchers.

Cuoco finished the morning forecast and began sending it by fax to BLM fire-dispatch offices in Grand Junction, Craig and Montrose in Colorado, and Moab, Utah. Once more he issued the highest alert.

"Red flag warning this afternoon and evening for hot and dry conditions," the forecast read. Increasing southwest winds would be troublesome for the many small fires started by lightning the day before.

"A high potential for large fire growth."

SUNDAY MORNING, AFTER full light, Jo Temple went to the second floor of her house overlooking Storm King and opened one of the big glass windows. The air carried a tinge of electricity, a "scary calm," as though last night's storm had left behind unfinished business. She saw nothing extraordinary, though, and settled down at a table where she did sewing.

At midmorning she glanced up and saw a wisp of smoke from Hell's Gate Point. Temple grabbed the telephone and dialed the Garfield County emergency center, which handles everything from treed cats to forest fires. They quickly relayed the report to the BLM's Grand Junction District office.

Others saw it, too. One man called in from South Canyon, a nondescript gully containing a few small ranches, old mine works and a landfill with a view of Storm King Mountain from the opposite side of the Colorado River and I-70. The man phoned directly to the BLM's Grand Junction district office and reached Flint Cheney, the lead dispatcher. He told Cheney he could see a "small whiff of smoke," but not its source. Cheney gave the smoke report the next fire number on his list, V891.

Minutes later Cheney received a call from Grand Junction's Western Slope Coordination Center. They had heard the smoke report and happened to have an aerial observer aloft in that vicinity. As part of normal procedure they needed a name for the fire so the pilot could check it out. What happened next was a small, unimportant mistake, but it set a pattern: In the days ahead, small mistakes became a series of unforgiving errors that finally compounded into one giant screwup.

Since Cheney had just heard from South Canyon, he used that name even though the fire was on the opposite side of the river. When asked why he didn't change it to the Storm King fire once he learned the true location, minutes later, Cheney's shoulders drooped. "It would have been more of a hassle to back out," he said.

To this day the events on Storm King Mountain in July of 1994 are known as the South Canyon fire.

BY LATE SUNDAY morning reports of the smoke on Storm King were coming in from every quarter. Members of the Glenwood Springs Fire Department, a mix of volunteer and paid firefighters, saw the smoke as they fought another fire nearby and called the Garfield County sheriff's office in Glenwood Springs. Sheriff's Deputy Steve Stebbing was dispatched, and became the first official on the scene.

Stebbing took I-70 to the foot of the mountain and watched as the fire spread from one to two trees. Alarmed, he radioed a request for the Glenwood Springs Fire Department to "extinguish the fire if they could." But they refused, saying the fire was in "very rugged" terrain and would take hours to reach.

The Glenwood Springs department had no legal and perhaps no moral responsibility to fight the fire, though afterward everyone wished there had been more cooperation all around. Following brief initial confusion, it had been established the fire was on BLM land and not private property. The main job of the Glenwood Springs department was protecting the city's homes and businesses, not making excursions onto federal property to fight wildfire, the universal term for fire in the open.

"We knew about the lightning strike," Jim Mason, director of emergency services for Glenwood Springs, said later. "Primarily our focus is for structure [homes and other buildings] firefighting."

Neither the Glenwood Springs department nor the New Castle Volunteer Fire Department, the next jurisdiction to the west of Storm King, had made its firefighters take the basic skills course and physical-fitness test required to fight wildfire on federal lands, though they had some wildfire training.

The BLM for its part was skeptical about the ability of any volunteer fire department to help them. In an overall fire-management plan completed in 1992, the BLM's Grand Junction District noted that county and volunteer fire departments had plenty of crews and equipment, but questioned their reliability.

"There is a concern for the dependability of their equipment, and the lack of fire behavior and other safety related training," the 1992 plan said.

All this did nothing to stop citizens watching the fire from bombarding the Glenwood Springs and New Castle departments with demands that they act, and in a hurry. So many spoke to Don Zordel, chief of the New Castle department, that at one point he drove to Storm King to look at the fire for himself. He pulled off I-70 at Canyon Creek Estates, estimated the fire to be a long mile beyond his jurisdictional boundary and drove home.

After the fire, Zordel and others in his department caused a stir with angry public remarks that the fire should have been put out sooner and that the New Castle department could have helped, statements widely interpreted to mean that the BLM had refused an offer of assistance.

"We could possibly have been of some assistance, yes," Zordel told *The Glenwood Post.* "This particular fire was not in my district, but the policy in my district, even if a fire is on federal land, is to put it out."

Discussing this in a later interview, Zordel acknowledged that he had never volunteered his department's services. In fact, he said, the BLM often helped him with fires. When he had made those remarks, he said, he had been blind with rage at the subsequent events on Storm King Mountain, ready to lash out at anyone. He was preparing now, he said in the interview, to have all members of his department qualified to fight wildland fire on federal lands, using training funds approved by the Colorado state legislature in the wake of the South Canyon fire.

It took until early afternoon Sunday, July 3, before any firefighter was dispatched to Storm King. It was nearing 2:30 P.M. when the loudspeaker outside the Western Slope Coordination Center in Grand Junction clicked and a metallic voice called out, "One load of jumpers!" The voice stirred action at the smoke-jumper ready shack located only a few feet from the Western Slope center.

Smoke jumpers are the elite cavalry of firefighting, parachuting onto small fires in the worst terrain the West can offer. They strike quickly, work fast and move on, leaving large, time-consuming fires to others. They live on adrenaline and $10,000 to $15,000 a summer in regular pay plus overtime, justifying their high training and transportation costs by keeping small blazes like the South Canyon fire from becoming big problems. In 1994 there were 388 smoke jumpers nationwide, the small number far out of proportion to the big job expected of them.

The ready shack at Western Slope, home for twenty jumpers, is located as far as possible from the passenger terminal at Walker Field, a circumstance the jumpers take personally and use to nurse a self-image as outside dogs, unloved until they become essential for the hunt. Waiting for a fire call, they lounge about like a pride of lions, restless, bored and a touch dangerous. One or two young men strut around shirtless, skin taut even around the belly and muscles cut in high relief. They wear their green pants rolled up—the fire-resistant material allows only poor ventilation, and rolled trousers add style. They keep their heavy logging boots on even in the desert heat—it takes too long to lace them once the alarm sounds. They know that the day's work has ended when someone calls out, "Let's take off the boots."

Ask how they're doing, and they grin and reply, "Living the dream!"

One wall of their ready shack stands open, pushed up like the door of a suburban garage. Inside, folding tables are strewn with a collection of well-thumbed magazines and paperbacks on sports and war. Two sewing machines stand in a corner ready to do repair work on the jumpers' elaborate rigging. A freezer holds the plastic water bottles that are handed to jumpers as they head for a fire, a welcome treat.

A single bathroom serves all. Women, who first broke into the jumper ranks in 1981, now make up about 5 percent of the force. The rear door of an airplane hangs from a wall hook like a poster, the only ornament in the place—jump planes fly with the door off. The pilots and air crews have their own house trailer with screen doors, air-conditioning and television; the jumpers get nothing so cushy.

Outside, a couple of stunted willows, which receive lavish watering by the jumpers, provide minimal shade. A chain-link fence surrounds the facilities, giving them the look of a minimum-security prison. The occasional coyote or rabbit can be seen beyond the steel mesh, loping through sagebrush.

When the loudspeaker announced the call for the South Canyon fire, George Steele, at the age of forty-six one of the most experienced men in jumping, made the short walk to the Western Slope center. He would act as spotter on this mission, picking the drop spot and providing the link between the jumpers, air crew and dispatch office. A tap from his hand would send jumpers into free fall.

Steele, his black hair beginning to grizzle, had been at the job so long he sounded like an action report, his speech studded with military jargon such as "affirmative" and "negative" for yes and no, and phrases like "we proceeded on." But the only thing Steele regretted about his chosen career was its mandatory retirement age of fifty-five.

Steele picked up a sheet giving the "lats and longs"—the latitude and longitude—for Storm King Mountain and an initial size-up of what they faced: "rugged, inaccessible terrain." Tailor-made for jumpers, Steele thought. The orders called for eight jumpers, one full planeload, plus a lead plane to act as "eye in the sky" and an air tanker with fire retardant, a red mud that slows flames.

Steele hoisted himself aboard the jumpers' plane, a light twin-engine affair, and took the copilot's seat. The jumpers filled the passenger area, stripped of its seats to accommodate them in their bulky jumpsuits. Aircraft J49 reported itself rolling for Storm King Mountain at 2:36 P.M.

Behind in quick succession came L64, the smaller lead plane, and T140, the air tanker, an outsize tail making it look like a lumbering balsa-wood glider. The call letters stood for each airplane's function: J for jumper, L for lead plane and T for tanker. The last was aloft by 2:44 P.M.

IT ALMOST SEEMED unfair to match this force against a couple of burning trees on Storm King Mountain; the three aircraft amounted to an armada of the skies, capable of striking several small fires, dropping as few as a pair of jumpers on each, or hitting a rolling blaze with everything they had. But it made good economic sense. Flying the air group cost in the thousands of dollars: An average of $407 per hour for lead and light jumper planes, a day rate of $1,500 to $2,800 for air tankers plus $1,200 to $3,000 an hour extra for flying time. By comparison, a runaway fire can cost a million dollars a day.

The jumper plane rose from the desert floor, crossed the Book Cliffs and flew up the Colorado River Valley, following the same path as the storm the day before. Only ten minutes into the flight, Steele, watching out the window, saw smoke and flames rising from the valley floor. Afternoon winds fanned a sagebrush fire, unreported until now. A dark gray thundercloud lurked nearby, the probable cause of the blaze.

Steele radioed the location to the BLM dispatch office in Grand Junction. The fire was "two acres burning in sage" about seventeen miles east-northeast of Grand Junction, he reported, in a place called Colbert Flats. The jumper plane continued on to Storm King, fifty-five air miles away, leaving the flaming sagebrush behind.

The air squadron followed the river and crossed the bristling hump of the Grand Hogback, still on the storm's old path. Ahead Steele saw Storm King rising over the V marking the Colorado River. With the plane about five miles from target, Steele began a pre-jump check, matching the terrain against a map. His eyes darted from the map to the view: Something wasn't right; he could see neither smoke nor fire.

A dazzling sunlight could be masking the smoke at that distance, he thought, or the fire could be hidden in one of the gullies. At most the fire would amount to a "good deal," meaning a night of overtime, a quick knockout, and off to the next good deal.

Steele didn't like the look of the mountain, though, even with no fire in view. The ridges were sharp, the slopes precipitous and the oak and PJ thick as dog's hair. "How am I supposed to find a jump spot in that shit?" he asked himself.

Steele called Grand Junction to report their arrival but never got beyond his first words. Dispatch interrupted with new orders: A disembodied voice recited a fresh set of lats and longs that for Steele carried a familiar ring. He checked, and sure enough, they were the same ones he had called in for the sagebrush fire; they were to turn around and go back the nearly fifty-five miles they had come.

Steele felt, if anything, a sense of relief: If dispatch thought the sagebrush fire more important, that was fine with him.

Only forty minutes elapsed between the time Steele first reported the sagebrush fire and the time the smoke jumpers returned to fight it, but in those minutes the wind had shown what it could do, doubling and tripling the size of the blaze. Gusts of wind hammered the jump plane as Steele, clutching an overhead cable, made his way to the rear. He tossed out streamers and watched as they flattened, edges fluttering, in the wind.

The thundercloud turned and swept back toward the fire, trailing strands of moisture. Steele figured that in ten minutes it would be on top of them. He motioned the jumpers to get ready, and the first, tightening parachute straps one last time, stepped forward and clutched the edges of the open door.

"On final!" Steele shouted to the jumper, the sound all but lost in the hundred-mile-an-hour wind and engine roar.

"Get ready!" Steele shouted, and the jumper rocked back.

Steele slapped the jumper's shoulder, not trusting words, and the jumper dove forward, curling into a tuck. The sudden smack of air can bring an involuntary gasp, like the first breath of life. The jumper counted to five, and the parachute snapped open, jerking him erect.

The jumpers went out in groups of two, called "sticks," the plane making a full circle, eating up more of the clock, for each stick. The thundercloud bore down; they had cargo yet to drop.

The pilot, Kevin Stalder, took the plane down to three hundred feet, and Steele began pushing and kicking out cardboard boxes filled with tools, food and drinking water. The low altitude and their small parachutes reduced drift, and the boxes smacked into the ground at the jump spot. As the plane made the last cargo pass, a distress call came over the radio: "We've got an injured jumper on the ground and need the emergency medical kit. Can you drop it?"

The thundercloud was close enough to darken the sky. Stalder, the pilot, figured he might squeak in one more run, with luck. He again dove to three hundred feet, and Steele readied the medical kit. As they swooped over the jump spot, the wind rocking the plane, Steele kicked it out the door.

"Hey, we got to get out of here!" Stalder called back on the intercom. "That's the last one."

The sagebrush fire spread to more than 160 acres by nightfall. Once the storm passed, a helicopter retrieved the injured jumper, who had a severely bruised heel from a bad landing. It took twenty-four hours and more than fifty firefighters—the jumpers, three fire-engine crews, two ground crews and an air tanker—to bring the sagebrush fire under control.

THE JUMP PLANE, J49, never flew close enough to Storm King Mountain to attract the notice of a group watching the fire there from the highway. Winslow Robertson, the number-two fire official for the BLM in the district, had driven the ninety miles from Grand Junction and arrived about 3:00 P.M., meeting up with a fire-engine crew from Rifle, only twenty miles away, who were already on the scene.

Robertson was known among firefighters as something of a "wallflower," a supervisor reluctant to commit force early to keep fires small. This worked well enough during a normal season, when fires often burned themselves out. Asked later if he ever took a hands-on approach, Robertson seemed startled by the suggestion. "I've never gone out and kind of, you know, strongarmed the situation and changed things," Robertson told fire investigators.

By the time Robertson showed up, the Rifle engine foreman, Clay Fowler, had already radioed a size-up of the fire to BLM dispatch in Grand Junction. Fowler said the fire was burning in "rugged, inaccessible terrain," and he would keep a watch on it rather than hike in.

That description—"rugged and inaccessible"—quickly became a universal rationale for inaction, repeated without question in official reports and in public utterances of responsible officials, though it frustrated homeowners and many others at the time. Jo Temple, who made numerous calls to inquire why the fire wasn't being fought, said later, "I remember the dispatch lady telling me it's 'inaccessible and we can't do anything about it.' I thought,

My God, one man with a shovel could have put it out the first day. The whole thing was not right."

Even with hindsight, the report of the official fire investigation team, called the South Canyon Fire Investigation, flatly stated, "The fire was inaccessible." Asked about this, Lester K. Rosenkrance, who headed the investigation, acknowledged that "inaccessible" was a misstatement and should have been amended to "inaccessible by [fire] engine."

Storm King Mountain was steep and covered with brush, but it could be climbed. Elk hunters roamed it each fall; an old Jeep trail led up its north slope; firefighters eventually would hike up its eastern slope, not once but two days in a row; and in years to come, tourists in tennis shoes would follow a new path up its western slope by the hundreds and thousands.

The fire on Sunday, July 3, was a matter of inconsequence except in its potential. It smouldered in a tree or two; it burned a few clumps of pine needles, a bit of dead grass, a fallen limb. This was an enormous advantage for firefighters if they had fought it then, as required by BLM policy for the area.

"Right now we are in a situation with severe conditions," Lynda Boody, assistant district manager of support services for the BLM's Grand Junction District, had written top district managers on June 14. Boody had overall administrative authority for fire operations, but deferred on matters of tactics and policy to Robert P. "Pete" Blume, the fire-management officer for the Grand Junction District. In the June 14 memo, Boody told the managers to attack, or "suppress," all fires rather than simply keep an eye on them.

"Weather has dictated Red Flag conditions for who knows how long," Boody wrote. "The fires we've had so far are burning into the night and are being exacerbated by high, dry winds. The point of this is that Pete [Blume] is inclined to take only suppression action while these conditions persist. That means we will not monitor fires but suppress them.

"Unless you have a reason, from now until the weather changes, the FMO [Blume, the fire-management officer] will take suppression action on all fires in this district. The bottom line is that we may be in for a long fire season this year."

The BLM also was required to spend as little money as possible fighting a fire. "Therefore," said a separate district policy memo, "we cannot place a fire in monitor status if the possibility is good that we will incur greater costs to suppress it at a later date." On July 3 and for several more days, the fire on Storm King would have cost a few thousand dollars to extinguish; by the time it actually burned out, it cost millions.

There was an additional policy specific to Storm King dictating immediate action. A "mandatory fire exclusion zone" had been established for portions of the mountain by the BLM's 1992 fire plan, the same one expressing skepticism about volunteer fire departments. All fires within this zone, drought or no drought, were to be fought on sight, on grounds they automatically threatened life and property, the highest priorities for fighting fire.

Storm King Mountain was surrounded on three sides by homes and other structures. In addition to Glenwood Springs to the east and Canyon Creek Estates to the west, a scattering of buildings lay to the south along the base of the mountain, including the BLM's local office in West Glenwood, an unincorporated community.

On July 3 the South Canyon fire burned in or just on the edge of the mandatory fire-exclusion zone. The South Canyon Fire Investigation says that the fire was inside the boundary. The BLM's Winslow Robertson later disputed this, saying that the fire on July 3 was just outside the zone. A map of the zone makes it appear that the ignition spot, Hell's Gate Point, is right on the boundary line. In any case, no one disagrees that the South Canyon fire later burned a substantial portion of the mandatory fire-exclusion zone and caused the evacuation of homes and businesses, including the BLM's West Glenwood office.

At the time Robertson took his first look at the fire, his boss, Pete Blume, believed that the Grand Junction District's firefighting capacity was virtually overwhelmed. Blume estimated that the district had forty fires burning at the same time, nearly half the average of seventy-nine for an entire year. Battling forty major fires simultaneously would have been a heroic if not impossible feat for six engine crews, the smoke jumpers at Walker Field and a handful of local crews, which was the most the BLM's district could count as its own. Blume later told fire investigators that he was spending his time at this point trying to find more crews and equipment.

"We couldn't get the number of air tankers we needed—air tankers, helicopters, crews, all of those things," Blume said.

The investigator asked, "Are you saying things became unmanageable?"

Blume responded, "I'm reluctant to use 'unmanageable,' but we were certainly taxed to the very limit. We do not have an organization that is big enough to run at that level."

The South Canyon Fire Investigation supports Blume's view. "The above normal fire activity overtaxed a relatively small firefighting operation," the report states, and now "overtaxed" joins the company of "inaccessible" as a shorthand explanation for what went wrong.

But was the Grand Junction District truly the heroic victim portrayed by Blume, faced by an overpowering situation that excused throwing out standing orders to suppress all fires and making halfhearted efforts with small, difficult ones?

And if things were that bad, or even close to it, why did the district have so much trouble getting help?

A closer look uncovers a significantly different picture, that of a small fire organization overwhelmed by a handful of blazes, not forty; its sense of emergency blunted by the demands of an already long and severe fire season; and its ability to shout for help drowned out by its own petty bickering.

Of the fifteen new fires reported on the BLM's Grand Junction District on July 3, only two were larger than five acres. One of those was the sagebrush fire attacked by Steele and his smoke jumpers; the other spread to sixty acres, driven by afternoon winds, and was handled by two fire engines and one crew of firefighters.

Three others were less than five acres, requiring only limited attention. The remaining ten were minor, a collection of unverified sightings of smoke and small fires that wound up being monitored from a distance, often by neighboring landowners, or being handled by other agencies.

In addition to the fifteen new fires, there were twenty-three old or "carryover" fires on the district on July 3. That brings the total for that day to thirty-eight, which can be fairly rounded off to forty as Blume did. But of the twenty-three old fires, the BLM actively fought only two. The larger of those had burned an impressive sixteen hundred acres but by July 3 was in "demobilization phase," meaning it was under control and firefighters were being released for other work.

The smaller one, a fifty-acre blaze, had twenty-three firefighters with four engines working on it. There had been enough progress, though, that the July 3 daily report predicted the release of some firefighters by the next evening, July 4, an estimate that later proved correct.

The remaining twenty-one old fires required almost no BLM attention: Eleven were under watch, mostly by non-BLM personnel; seven were listed as dead or under control; and two were unverified reports of smoke. "No smoke reported," reads the summary for fire number V888, and "No smoke spotted" for V886. The summary for V883: "Dispatch received a report that the fire was acting up. Engine 674 and a Utah engine were dispatch [sic] to the area but could not located [sic] anything."

The summary for V891, the South Canyon fire, puts the size of the blaze on July 3 at two tenths of an acre. It notes that an engine crew was dispatched. "The fire was inaccessible and had low spread potential," the report says, picking up the "inaccessible" label.

Then a faint but audible note of skepticism slips in. "It was very visible to I-70 but was placed in monitor status," the daily report says. Preventing a public-relations nightmare may not be the best reason to fight fire, but it is common to try harder with fires in sight of towns and major highways, and the person writing the log apparently thought so, too.

The South Canyon fire, for its part, played an almost deliberate game of hide-and-seek. It smoked enough in its two-tree base to alarm motorists and nearby residents. But when the proper authorities showed up, it disappeared, first when the jumpers came in sight, and then a second time as Winslow Robertson and Clay Fowler watched through binoculars.

In late afternoon on July 3, the skies over Storm King darkened and a thunderstorm rolled in with the welcome scent of rain. This storm carried seventeen hundred lightning strikes, not as spectacular as the day before but enough to seed many new fires. Rain was a rare treat for Glenwood Springs, a center of the drought where no rain had fallen since June 22. Precipitation in the town had been less than half normal for the months of May and June, a total of 1.16 inches compared to an average of 2.79 inches.

Robertson took a look at the clouds and decided to call it a day. He felt miffed that he had shown up at all; he drove to the fire, he said later, only because the Garfield County sheriff's office had "made it sound so extreme." He figured he had done all he could: dispatched an engine, conferred with the engine foreman and made the judgment that it would take too long to hike in. The fire, being on top of a ridge, looked as if it had nowhere to go.

"When I looked at it, it was so small," Robertson said later, sounding plaintive. "I didn't really think it was an issue."

Fowler stayed at the scene for another quarter of an hour. A few drops of rain began to fall, enough to tamp down flames. The smoke disappeared from Hell's Gate Point. Fowler got in his truck and drove off.

TWO

WHEN CHRIS CUOCO arrived at work on Monday, July 4, and checked the overnight weather data, he discovered a cold front building in northern Idaho and northwestern Montana. Cuoco figured it would arrive in Colorado in two days and with luck would bring enough rain to ease the drought. But its advancing edge would carry heavy, shifting winds, more bad news for the beleaguered fire community.

The Fourth of July was turning into another hot, blowy day with afternoon thunderstorms, and Cuoco prepared his second Red Flag Warning in a row.

Holiday or not, the day promised to be busy; the governor of Colorado, Roy Romer, was visiting fires across the state to show personal concern. Romer was due at the center at midmorning, and Cuoco was to give him a fire-weather briefing. Romer, a veteran politician with a down-to-earth manner voters liked, faced an unexpectedly stiff reelection challenge from Bruce Benson, a wealthy Republican businessman. The issue was money.

Benson was attacking Romer for failed public-spending projects, notably the Denver International Airport, at the time a financial and engineering disaster. With fire-suppression costs sky-rocketing, Romer acted like a man on a tightrope, trying to strike a balance between the threat to life, property and forests and the need to control spending. He had become a fixture on the evening news, complete with a yellow fire shirt and a white hard hat.

At 10:00 A.M. the governor was promptly ushered to the pilot's lounge, where Cuoco, whose years with NORAD had polished his briefing manner, gave him a weather report with a warning about the next two days. Rain may provide relief later in the week, Cuoco said, but "first we'll have to get through Tuesday and Wednesday."

The thunder of an aircraft engine interrupted the briefing, and Romer, a pilot himself, jumped up and looked out a window. A converted B-24 Liberator bomber was lumbering along the tarmac, followed by more heavy tankers; something big was afoot.

The briefing broke up. The governor met with reporters and quoted from Cuoco's forecast, while Cuoco headed back to his cubbyhole, a windowless utility room with a folding table and one chair. En route, Cuoco heard Mike Lowry, who had helped brief the governor, on the telephone in another office.

"We've got to get some people here; we need more resources!" Lowry barked into the telephone, and then added with a note of despair, "But I don't know how much chance we have of getting this stuff."

SHORT, BRISTLY AND energetic, Lowry was a thirty-year fire veteran who, like Cuoco, had been called to the Western Slope Coordination Center a few days earlier to help with the emergency situation. Lowry, whose regular post was assistant manager at the Northwest Coordination Center in Portland, responsible for fire suppression in the timber-rich states of Oregon and Washington, was running operations at Western Slope while the manager, Paul Hefner, handled administration.

From the moment he arrived, Lowry had seen disaster in the making.

Cooperation, the touchstone of modern firefighting, was virtually nonexistent. Instead, Lowry found competition, jealousies and outdated policies and thinking. A feud smoldered between the BLM's Grand Junction District and the Western Slope Co-ordination Center, located only a mile apart. Personnel from the two offices hardly communicated, acting more like dysfunctional children than cooperating adults. Lowry could see a fleet of air tankers under Western Slope command sitting idle at Walker Field each morning while it was cool and windless, the best time to fight fire. But strictly speaking it was the Grand Junction District's job, and not Western Slope's, to request tankers, and they held off each day until afternoon, saving the cost of a tanker but risking a runaway blaze. By then sun and wind had kicked new life into old fires.

BLM fire officials at the state level were no more effective. Don Lotvedt, who held the top fire job in the state for the BLM, State of Colorado Fire Management Officer, had no fire experience. He had been placed in that post after his previous job in an unrelated field was eliminated by downsizing. His subordinates tried to compensate for Lotvedt's inexperience, but he occupied a key position.

The BLM's director for Colorado, Bob Moore, stayed aloof from fire work, later acknowledging to fire investigators that he and his top managers did not become involved until after "obvious" problems had emerged. Under Moore, Colorado, virtually alone in the fire world, held on to a policy of allowing no air tanker to drop retardant unless a crew was on the ground to follow up. That policy had been abandoned elsewhere, as air tankers working alone proved they could fight many small blazes at the same time, knocking out the smallest and wetting others until firefighters arrived. The ideal situation for a single tanker was a fire in a tree or two in difficult terrain during a busy fire season, a description that fit the South Canyon fire on July 3.

Worst of all, nobody at the BLM was calling for help.

"It was very frustrating to see the way they were doing business," Lowry said later. "We knew what was going to happen. We knew we were into more lightning; it was getting drier and drier; we *knew* the fires were going to be larger.

"And we just didn't have anything to fight with!"

Lowry's job in Portland was to prevent exactly this situation from developing. When a region ran short of crews and equipment, it was a coordinator's job, supported by the National Interagency Fire Center in Boise, to locate what was needed and fly it in. That meant anticipating trouble and having crews and equipment in place before a crisis hit.

When Lowry arrived, it was taking a full twenty-four hours for crews to be flown in to fight existing fires, never mind the threat of fires in the future.

"We were really behind the curve," Hefner, the manager of the Western Slope center, acknowledged later.

Lowry set out to change that by ordering a massive number of firefighters—twenty crews, or four hundred people—to be flown in immediately and held in readiness around Grand Junction. The order was supposed to go to the Rocky Mountain Coordination Center just outside Denver, responsible for all of Colorado. But nothing happened.

"I got into several conversations with the coordinator in Denver about moving crews and overhead and pre-positioning in Grand Junction. They were just not aware of how to go about it. I don't think anybody in that office had been required to do that.

"They were extremely upset. They would not place the crew orders the first couple of times we tried," Lowry said.

David Clement, area coordinator for the Rocky Mountain center in 1994, said he and Lowry spoke "continuously" about conditions in western Colorado in June and early July. Clement said he was sending out crews on a daily basis during that time, mainly to fires in eastern Colorado. When fires became more common in western Colorado, he said, he sent additional crews there. The Rocky Mountain center also covers Kansas, eastern Wyoming, Nebraska and South Dakota, but those states have few fires compared to Colorado.

Clement denied refusing to forward any of Lowry's requests, and said he could remember no "heated" conversations with him. "Oh, no, nothing like that," said Clement. "Things kept escalating, it was a continuous buildup." The result, however, was too few crews pre-positioned in western Colorado in the first days of July. Clement subsequently was transferred to a position with less responsibility, assistant for fire operations.

By 1994, fighting wildfire had come a long way from the days when a ranger recruited a fire crew by clearing out the local saloons. Spurred by the needs of an expanding population dependent on fire control and by budget cutbacks and consolidation—the cost of fighting wildland fire in an average year is a substantial $1 billion—the federal government, starting in the 1960s, began to blend together the various agencies responsible for fire. It was no easy task and even today remains far from complete.

The Forest Service, which had long reigned supreme in wildland firefighting, struggled to cooperate with rival agencies, the BLM in particular. The BLM's acreage, the land "nobody but God and the BLM could love," receives less public attention, being the leftovers from the Homestead Act, the Forest Service, National Park Service and other agencies. But the BLM is the nation's number-one landholder—268 million acres, including most public lands in Alaska, compared to 191 million acres for the number-two landholder, the Forest Service.

Since the 1960s the federal government has built its own summer army of trained firefighters, numbering about ten thousand. These range from migrant farmworkers on emergency call to the highly trained smoke jumpers and "hotshots," ground crews of mostly college-age youngsters who fight the bigger fires. The emblem of fire season in the West has become a row of school buses loaded with youthful firefighters, their ashen faces nodding in exhaustion above grimy yellow shirts.

At the time of the South Canyon fire, the headquarters for the firefighting system had just acquired a new name—the National Interagency Fire Center—and a new brick-and-glass administrative building, finished only that April. The center, known as NIFC (pronounced NIF-see) is at the Boise Airport, giving Boise a claim to be the nation's capital for fighting wildland fire. NIFC was known before 1993 first as the Great Basin Fire Center and then as the Boise Interagency Fire Center as it evolved from regional to national responsibilities. The NIFC site in Boise,

owned and managed by the BLM, hosts six federal agencies con-
cerned with fire: the BLM, Forest Service, National Park Service,
Fish and Wildlife Service, Bureau of Indian Affairs and National
Weather Service.

As monitor of the national fire situation, NIFC commands the
smoke jumpers and hotshots, known as Type I crews, as well as
teams of fire supervisors and the heaviest aircraft, the tankers and
biggest helicopters. Together they are known as "national resources" and are first to be dispatched across regional boundaries.
In bad fire seasons such as 1988, when fires swept Yellowstone
National Park, the federal government can summon the U.S. military to help on the line. The United States even has mutual-
cooperation agreements with Canadian provinces. Once assigned
by NIFC, personnel and equipment come under local control as
they engage individual fires.

Brochures proclaim NIFC the fulfillment of "interagency cooperation at its best," but in 1994, as well as today, that amounted
to overstatement.

NIFC occupies a sprawling complex, a city or campus unto
itself. The NIFC warehouse has the largest cache of fire equipment in the country—its catalog runs to 222 pages; its radio shop
services thousands of handheld radios, telephones and other communications gear; its machine shop develops new fire engines
from scratch; and it has offices for infrared mapping, aerial im-
agery, weather-data collection and other technical services.

The buildings include a rectangular tower, the distinctive structure of every smoke-jumper base, used for hanging parachutes.
Since 1986 NIFC has been home for the Great Basin smoke
jumpers, named for the vast inland basin that stretches from Nevada to Idaho. They are one of two smoke-jumper units who
wear BLM and not Forest Service uniforms, the other being in
Fairbanks, Alaska.

During fire season the day begins with fire and weather briefings for key personnel. A stack of situation reports summarizes
large fires in the United States, and a single brusque sentence can
speak volumes: "Alaska: Ten fires are in limited protection status
totaling 342,723 acres [over 535 square miles]." A weather forecaster, using video techniques worthy of network television,

points to blue and red weather fronts moving across a screen, predicting which will cause trouble for firefighters.

The heart of NIFC is the coordination center, a cross between a war room and an air-traffic-control center, located in the headquarters building. Aircraft locations are tracked on a wall map in a large, open room. Incoming calls are answered at a raised dais, then parceled out to four separate stations, each with desks and computers arranged in a star shape for ease of communication. The desks handle supplies, air and smoke-jumper operations, crews and overhead, and intelligence or situation reports.

The resource-ordering system works from the bottom to the top; anyone from a radio dispatcher on up can order from the next higher level. Requests come in to NIFC from eleven co-ordination centers across the country, including the Rocky Mountain center near Denver and the Northwest center in Portland (the Western Slope center was a subdivision of Rocky Mountain). In daily conference calls, coordinators talk with NIFC about present and upcoming needs.

One of the most difficult decisions NIFC faces is whether to commit crews and equipment before an outbreak of fire, when there is only a threat. This is exactly the situation Mike Lowry at the Western Slope center and David Clement at the Rocky Mountain center faced in western Colorado in early July. The procedure for commitment, which so baffled Clement, has its own name at NIFC—"move up and cover." When the fires actually break out and crews are already on hand, everyone looks like a genius; when the threat fails to materialize, the most valuable crews and equipment wind up stranded.

Recognizing a need for fast, emphatic answers, NIFC has an hour to "fill or kill" a request after it is made. "It happens really quickly," said Kim Christensen, NIFC's assistant center manager. "When a request for an air tanker comes in here, it takes one to three minutes to pass along the information. Within fifteen minutes the props are turning . . . that's how fast it *should* happen."

In short, NIFC is an up-to-date technological wonderland combining weather satellites, computer networks and shorthand terms for complex decisions. But as often happens with the elec-

tronic revolution, and with snappy jargon, matters seldom come
out right unless an informed human intelligence makes the correct judgment at the proper time and, in fairness, has a bit of
luck.

ON JULY 4, Independence Day, Lowry was determined that his
crew orders would be honored by the Rocky Mountain Coordination Center. The lightning storm on July 2 had given him
the argument he needed—a fresh outbreak of fires—but he was
taking no chances. He had made a round of telephone calls to
contacts in the national firefighting system to line up political
support.

The crosstown rivalry between Blume from the BLM district
and Hefner from Western Slope had a long history, notorious in
Colorado and beyond, and Lowry had no false hope of resolving
it. The time had come, though, to insist that they open more
direct communications, especially when it came to the crucial
matter of ordering crews and equipment.

Instead of writing requests and trusting in goodwill, the BLM's
Pete Blume and his assistant, Winslow Robertson, had taken to
playing cat and mouse with Western Slope rivals. When they
needed an air tanker or lead plane, Robertson would drive—or
in his words "bop on over"—to Walker Field to see what airplanes were standing on the tarmac. Then he would drive back
to the district office and request those specific planes, knowing
he couldn't be refused. Blume defended the practice later by telling fire investigators, "You don't ask for resources you don't
think you'll get." The game defeated the purpose of the national
system, which was to move crews and equipment by need, regardless of where they were stationed.

The BLM Grand Junction District officials had their own complaints about the Western Slope center, namely that it "hoarded"
resources and failed to tell them in timely fashion the disposition
of their requests.

"There was a very definite lack of trust. It surprised me; they're
neighbors, for God's sake," Lowry said later. "I spent quite some
time convincing Pete we were not going to screw him" by steal-
ing authority, airplanes, helicopters, crews or anything else. His

message to Blume and Hefner was the same: Start talking to each other.

The shift log, written from Hefner's point of view, records a success: "Saw Mike and I had a meeting w/Pete and discussed crews, ordering channels, etc. Came to agreement on process and he will try and meet w/me daily to touch bases."

Lowry had a less rosy assessment of Colorado's top firefighters: "Leadership in this state sucks," he later told investigators.

The BLM–Western Slope rivalry dated to 1978, when the Western Slope Coordination Center was established to coordinate air support among eleven small districts on the western slope of the Rocky Mountains. Conflict inevitably arose with the BLM's Grand Junction District, which had overlapping jurisdictions and functions. General personnel cutbacks in the 1980s and '90s aggravated the existing jealousies.

Part of the problem was the state of Colorado's reputation, perhaps undeserved, for being a firefighting backwater. It was true that the state had fewer big fires than the northern Rocky Mountain States with their vast pine forests, but its smaller brushfires could be spectacular and dangerous. In any event, fire management in Colorado, especially on BLM wastelands, was widely considered a dead-end job. Nobody's attitude improves under those conditions.

By 1989 a regular BLM audit of fire-safety procedures in the Grand Junction District found "serious morale problems." When more than one agency fought a fire, the result was "confusion and delays of the suppression effort," the auditors reported. As a result, "Efforts are inefficient, safety is compromised, and the public can be confused."

The next regular audit, by a team of BLM officials from outside the district, came in July 1993, a year before the South Canyon fire. This time the auditors found overall fire management in jeopardy. In one conclusion the auditors seemed to be writing a preview of the South Canyon fire: "Differences in resource management philosophies, personalities, misconceptions about the use of prescribed fire, and an unclear understanding of position roles and responsibilities seem to have created a difficult situation with respect to the management of fire."

The report cited Grand Junction managers specifically for allowing too many fires to burn unchecked. The managers "expressed a willingness to take more risk" with some fires than was acceptable, the report stated. It also noted that an employee in Glenwood Springs had complained that his office was cut out of the information loop, a gap that came to delay action in the South Canyon fire.

"Frequently they [BLM personnel in Glenwood Springs] learn of fires from other sources," the report said. "There is a perception that the [BLM's Grand Junction] District Office fire personnel are sometimes 'controlling all the shots.' "

The auditors called on everyone concerned to "work together to reconcile differences and come to terms with an acceptable approach to fire management," the same plea Lowry made to Blume and Hefner.

By 1994 there was, in fact, little reason for the continued existence of the Western Slope Coordination Center. Its functions now clearly belonged in the Rocky Mountain Coordination Center, but like every bureaucratic entity, it died hard. The center manager, Hefner, had himself drawn up a plan to modernize the operation, though it had more than a touch of empire-building. Hefner proposed that the Western Slope center become the region's coordination center, replacing Rocky Mountain. If not that, Hefner's second choice was to have Western Slope "incorporate" or take over dispatch services for the BLM's Grand Junction District and for the Forest Service's White River National Forest, headquartered in Glenwood Springs. At the time of the South Canyon fire, Hefner's plan languished in a file drawer.

None of the audits, plans or reports, however, touched on the sorest point in the Western Slope–BLM district battle: control over the one helicopter crew assigned to the region. Separate helicopter units at the BLM district and the Western Slope center had been consolidated in a cutback in 1992, with Western Slope given control of the surviving unit. Two years later the loss still rankled both sides, contributing to general ill will.

Hefner's attempt to smooth the transition in 1992 by naming the two existing helicopter foremen as "co-foremen" of the new crew only added a new level of tension. One of the co-foremen, Richard Tyler, started keeping notes of grievances. In the fall of 1993 Tyler put his notes together in a letter to Hefner, calling on him to resolve the situation by choosing between Tyler and his co-foreman, Patrick Basch.

"This letter is a formal statement and request for resolution to the problem which I have identified and discussed with you through the past two years," Tyler wrote. "For this problem to go unresolved will only mean a continued and increased safety risk of our helicopter program." On advice of friends, Tyler presented his arguments in person to Hefner, with Basch present, and never mailed the letter.

Tyler, one of five brothers all of whom were Eagle Scouts, was known as an up-and-comer with an extra concern for safety. He had been on a helicopter crew in 1986 assigned to a fire in rough country southeast of Grand Junction. At the last minute Tyler switched with another crewman and wound up driving a truck instead of riding the helicopter. Returning from the fire, the helicopter swung low into a canyon dark with shadows and ran full-tilt into high-voltage power lines. The crash killed the pilot and three crewmen. In the aftermath Tyler was promoted to helicopter foreman. "After losing his co-workers, his mission was always one of safety first," Patty Tyler said of her husband.

In the winter of 1994, Tyler was named sole foreman of the Western Slope helicopter crew, in time for the South Canyon fire.

BY EARLY AFTERNOON on the Fourth the rumble of aircraft at Walker Field had become nearly continuous as planes took off and returned for more fuel and supplies. There were five new fires on the Grand Junction District alone, one of which threatened ranch buildings. Another had burned out a set of high-voltage electric-transmission lines, cutting off power to thousands.

But the worst fire by far in the region was one endangering homes and lives in sagebrush country about fifty miles southwest of Storm King Mountain near the town of Paonia, outside the Grand Junction District in the neighboring Montrose District.

The first engine crew on the scene had been unable to find the fire, it was said later, and had started to drive off when one crew member saw a thread of smoke in the truck's rearview mirror. Before the crew could act, the wind caught the fire and sent it roaring toward homes scattered among the sagebrush.

Firefighters pulled out the stops for what came to be known as the Wake fire; Storm King would have to wait. At 1:49 P.M. the Western Slope center issued a blanket order: "Divert all tankers to Wake fire," and moments later the loudspeaker outside the center crackled, "One load of jumpers!"

George Steele began suiting up. He had made it back onto the jump list that morning, taking boyish pleasure in switching another jumper to spotter, a process known as "boning your buddy." By 2:01 P.M. Steele and seven other jumpers were aloft, followed within minutes by a second planeload of jumpers, which cleared out the ready shack.

As the first plane, with Steele aboard, came in sight of the Wake fire, Steele braced in the open door. A wave of heat boiled up from the fire, driving him back. He had seen enough, though: One home already was in flames.

The spotter picked a jump spot protected from the fire by an irrigation ditch and slapped Steele on the shoulder. As first jumper out the door, Steele automatically became smoke jumper in charge, a title that can mean a little or a lot. It turned out that Steele was senior to anyone on the ground and upon landing took charge of operations with the additional title of incident commander; the fire world has adopted the national Incident Command System, which gives a common organizational structure to emergency situations ranging from fires to hurricanes— the old "fire boss" has become an IC or incident commander.

While Steele was jumping, the fire had slipped across the irrigation ditch, and the home he had seen burning from the air had become a smoking ash pit. Sheriff's deputies began evacuating residents from the area. Four air tankers dropped red mud on other homes nearby, drenching one moments before flames reached it. The pilots, flying blind through billowing smoke, were too late for two other homes.

Steele's most immediate problem was the fire on his side of

the irrigation ditch. He set his jumpers, an engine crew and several bulldozers to work digging a line to check it.

The loss of the three homes drew Governor Romer and his entourage to the scene by late afternoon. With television cameras rolling, Romer gave his assessment:

"All hell broke loose. We have evacuated ten families, and we've got six to twelve homes being threatened. It's a very hot fire, very volatile. We're bringing in some more [help] from out of state."

Asked about the financial cost, Romer said that the state's firefighting fund was exhausted, but he had requested emergency funds from the federal government.

"We're going to put out the fire," he promised. "We'll worry about the money later."

ON STORM KING Mountain the fire revived on the morning of the Fourth and was never again a will-o'-the-wisp. It spread off Hell's Gate Point with the slow determination of lava. It sputtered, smoked and grew by inches, but by noon it would be three acres, and by nightfall more than ten.

The holiday edition of *The Glenwood Post* carried an unintended reminder that mountains like Storm King are made to be climbed. A front-page photograph showed two native sons, Steve Kibler and Mike Paddock, atop a ridge in Glenwood Canyon near town, the same perspective above the Colorado River and I-70 as from Hell's Gate Point a few miles downriver. The two men had been making the climb every year since 1968 to celebrate Independence Day, raising an American flag from the summit.

And once again on the Fourth no air-tanker flights took off with the dawn from Walker Field, though a substantial air force had been ready at that hour. Tankers normally are available at 9:00 A.M., and earlier by arrangement. On July 4 the Western Slope Coordination Center had command of five air tankers, two lead planes, a helicopter, two smoke-jumper planes and two additional air-attack planes. All wound up committed on the Wake fire, but not until afternoon. Similar numbers of aircraft were available on the days before and after the Fourth.

"If they [the BLM's Grand Junction District] don't order the air tanker, the air tanker doesn't go," Hefner said later. "They would wait until the smoke started, and then the air tankers would go nuts. We would send air tankers if they were ordered, but they weren't."

The South Canyon Fire Investigation notes this situation in a single sentence, pregnant with unstated conclusions: "Apparently there were intermittent opportunities where additional air support was available on July 3, 4 and 5, but they were not used on the South Canyon fire."

Why were they not used? Would they have been able to put out the fire if they had been called to action? Was anyone negligent in not sending them? The investigation report never addresses these questions.

The lack of action exhausted whatever patience was left in the community around Storm King Mountain. By midafternoon, the Garfield County sheriff's office was receiving so many telephone complaints that they began keeping count, logging more than two hundred during one twenty-four-hour period.

At 3:00 P.M. the BLM's Grand Junction District took its first action of the day on the fire: They tried to turn it over to the Forest Service. In response the service's White River National Forest, headquartered in Glenwood Springs, dispatched Sam Schroeder, one of their engine foremen, and his crew of two from the town of Rifle.

Schroeder took one look at the blaze and called the BLM's dispatch office in Grand Junction to request an air tanker. The dispatcher told him they had "no aircraft available" and asked if the fire was threatening any homes or other buildings. No, Schroeder had to admit, the flames were more than a mile from the nearest buildings. The dispatcher said they were sending tankers only to fires where "structures were threatened."

Schroeder and his two crewmen, Nicholas Strohmeyer and William Tarallo, shouldered chain saws and picked up hand tools. It was 4:00 P.M. on July 4, nearly forty-eight hours after the South Canyon fire had started, when they stepped forward, the first to take on the blaze. They never reached it.

Judging Hell's Gate Point too steep for a direct climb, they started up one ridgeline to the west, a gentler slope where today a tourist path begins. There was no trail in 1994, and the three struggled through PJ so dense they had trouble keeping the fire in sight. They tied brightly colored flagging tape to branches as a guide for retracing their steps. After a half hour they ran into loose footing on a steep slope and halted to take stock.

They could see down to I-70. As they watched, another fire engine pulled off at the Canyon Creek exit and parked. They called on the radio and raised James R. "Butch" Blanco, a BLM engine foreman and a native of Glenwood Springs who knew the area well.

Blanco told them that if they were stuck to come on down and talk things over. As they started back, Blanco made his own assessment of the fire and concluded, as had Schroeder, that an air tanker would be a big help. Like Schroeder, Blanco radioed the BLM's Grand Junction dispatch office and asked for one. He was told that he was number nineteen on the request list, likely an exaggeration, and that there was no hope of a tanker before dark.

Schroeder and his crew hooked up with Blanco, and together they scanned the hill with binoculars for a better route. While they were looking, yet another BLM truck pulled off I-70, a "six-pack" or crew truck capable of carrying six people.

Michelle Ryerson, a former flight attendant and now a BLM squad boss, arrived with a mixed crew: two Forest Service veterans and two "EFFers"—emergency, call-when-needed firefighters with almost no training. Ryerson had what is politely called a "personal situation" with one member of her crew, Loren Paulson of the Forest Service, who was an ex-boyfriend; after initial strain upon finding themselves on the same crew, beginning a few days earlier, they had settled down to a working relationship. In the coming hours they would share an intimacy more intense in its way than any they had known in three years of dating.

Blanco, Schroeder and Ryerson conferred and agreed that it would be dark before they could hike to the fire, and they had no camping gear for an overnight stay on the mountain. Besides, engine crews normally expect to remain within a hose's length of their trucks, versatile and expensive machines not intended to be left behind while those who staff them hike to remote fires.

As they talked, an observer airplane, sent by the Forest Service, came into sight. They listened in as the pilot reported on the radio, describing the fire as burning on the ridgetop "with some slop over on east side." A helicopter with a water bucket would be an effective tool, the pilot said, but cautioned that any fire-fighters should wait until morning before climbing the mountain because it was "too steep" for night assault.

This report became garbled as it passed from the Forest Service to the BLM dispatch office in Grand Junction. At 10:09 P.M., more than three hours after the observation flight, a BLM dispatcher logged in a far different version, one that used the now-familiar description "inaccessible" and recommended against fighting the fire under any circumstances. This entry reads:

"Aerial observer for Glenwood Springs flew fire. Steep inaccessible terrain, burning to NE to ridge, 2 steep for crew—suggest bucket drops, little or few escape routes—currently spreading in all directions."

That entry, edited to make it read more smoothly and in the process becoming more adamantly wrong, appears in the South Canyon Fire Investigation, adding to the picture of a mountain too rugged to climb, a fire too tough to fight, in keeping with the accepted view. "The fire is in steep and inaccessible terrain," reads the investigation entry, which appears in quotation marks as though a verbatim account from the dispatch log. "The area is too steep for crews and has few if any escape routes. The fire is actively burning in all directions. Helicopters with buckets could be very effective."

Fire investigators eventually reinterviewed the pilot, who confirmed, contrary to the later versions of his flight, that he had reported that the fire could be fought in the morning.

Blanco, Schroeder and Ryerson decided to call it a night. The fire "wasn't doing anything," Blanco said later, "just creeping around." The three crews pulled out shortly after 6:30 P.M. and drove to the BLM's office in West Glenwood, gathering in a cavernous shop at the rear where Blanco had a desk and telephone.

Blanco was an institution in the valley, a volunteer fireman in Glenwood Springs for twenty-five years before joining the BLM in 1989. At fifty, he was older than anyone else on Storm King, a man Pete Blume and Winslow Robertson had come to count on for handling their problems in the field. Blanco had a knack for putting people at ease and letting them do their job. He had completed a course in the spring of 1994 to qualify as an incident commander; the authorities waived the field-exercise requirements as a courtesy, a bending of the rules extended as a matter of course to those with long fire experience. This later led to a major misunderstanding, a rumor that Blanco was not qualified to serve as IC on the South Canyon fire. In fact, he was an officially certified IC. The rule-bending became controversial, but that dispute had nothing to do with Blanco personally.

From his desk Blanco telephoned Flint Cheney, the BLM lead dispatcher in Grand Junction, and asked about the possibility of more help in the morning. (By coincidence, Cheney and Blanco had been classmates in the incident-commander course, and Cheney like Blanco had been excused from the field requirements.) Blanco said he needed an air tanker and at least one helicopter. A helicopter could haul crews and equipment, saving the long pack-in, and make drops with a water bucket.

From listening to Blanco's end of the conversation, the crew could tell that Cheney had asked about a threat to homes and buildings. Somebody joked, "Can't we find an orphanage somewhere on that mountain?" But Blanco had to tell Cheney that the only structures they had seen were telephone poles.

Blanco pleaded his case, almost begging. Then he listened for a few moments. "I understand," he said finally. "We'll do the best we can."

It was about 10:00 P.M. by the time Blanco finished his call, and there wasn't much left to do except talk over prospects for the morning. The supervisors, Blanco, Schroeder and Ryerson, figured they could at the least count on their own three crews, ten people in all, though that turned out to be overly optimistic. The rest was up to chance. They might pick up a local crew with twenty more fighters. Blanco and his engine helper or partner, Bradley Jan Haugh, had been released from the fifty-acre fire that was coming under control, and another two-person crew might be available from that fire in the morning. With luck they could patch together a group of thirty or more hands, enough to mount a respectable attack.

They agreed to meet again at the BLM office at first light. The locals would spend the night at home, and the rest would sleep on cots in the shop. Schroeder left with his crew to check in with the Forest Service at the White River National Forest headquarters on the other side of town.

When Schroeder arrived there, shortly after 11:00 P.M., he was told that two more fires had broken out along the river and he had been assigned to one for the morning. The South Canyon fire belonged to the BLM once again and to Butch Blanco.

Ryerson and her crew watched the Glenwood Springs municipal fireworks display from the BLM's parking lot and then bedded down around midnight. They spent a fitful night. Each time they fell asleep, the telephone rang on Blanco's desk. In the dark the residents of Canyon Creek Estates could see flames out their picture windows, and they telephoned the BLM office for an explanation. One resident, a sleepless Allen Bell, went to his backyard on the southern edge of Canyon Creek Estates with his home-movie camera.

"July 4, 1994," said Bell, beginning the remarkable account he would continue for two more days. He could see the peak of Storm King and most of Hell's Gate Ridge. A low ridge, the same one Schroeder and his crew had hiked that afternoon, intervened between the Estates and Hell's Gate, about a mile and a half away, hiding the lower portion of Hell's Gate Ridge. Bell guessed he could see fifteen acres burning. A necklace of flames draped the ridge and outlined the lower edge of the fire. Occasionally a tree torched and sent branches spinning downhill, extending the fire's reach.

"If this was a big campground, these would all be campfires," Bell said of the flames. "I don't know what the other side of the mountain looks like, but it's on fire, too."

Bell panned his camera from Hell's Gate Point along the ridge-top about a mile to the place where it joins the peak of Storm King. He had to angle his camera up; he was at fifty-eight hundred feet and Hell's Gate Point was at seven thousand feet. Near the peak a dip, covered with scrub oak, formed a saddle or natural funnel.

The danger seemed obvious to him. "If it gets into that," Bell said, holding the camera on the dip, "there's nothing left but a nine-thousand-foot mountain to burn."

IT WOULD BE foolish to think that the conclusion of events on July 4 came about solely due to Mike Lowry's efforts, or Governor Romer's presence in western Colorado, or the loss of three homes on the Wake fire, though all played a role. At 6:15 P.M. the national interagency firefighting system, from Grand Junction to Denver and to Boise and beyond, finally focused on western Colorado. The log of the Western Slope Coordination Center records the moment, with a sly reminder that the Wake fire proved a good lever for getting more help for fires throughout the region: "We ordered 20 extra crews on the Wake fire and will disperse them *as needed* [emphasis added] when they arrive."

Overnight the far-reaching network required to airlift people by the thousands, earthmoving machines by the score and equipment by the ton began to churn. The first twenty crews became a vanguard, followed by multiplying numbers of management teams, smoke jumpers, hotshots and other ground help; air tankers and helicopters; bulldozers, backhoes, fire engines, water trucks, crew buses, shower stalls and chow lines.

Purchasing agents, food caterers, police and postal authorities would set up shop. Orders would flow for cases of disposable paper sleeping bags, men's and women's fire pants, gloves, fire shirts, chin straps and first-aid equipment—three hundred tins of lip balm, four hundred rolls of antacid tablets, a hundred eye patches, fifty biohazard bags and sixty large abdominal bandages.

Warehouses across the West would give up stacks of chain saws with repair kits and earplugs; shovels, Pulaskis, and other grubbing tools; tent poles, stakes and flies; fax and copying machines; hundreds of radios and thousands of batteries; and all the paraphernalia that in forty-eight hours of good flying weather can create an instant city.

An army was on the move.

THREE

BUTCH BLANCO AWAKENED after a few hours' sleep at home on Tuesday, July 5, and telephoned the BLM's Grand Junction dispatch office to renew his request for help. Finding no supervisors at work at that hour, 4:00 A.M., he left a message asking for a helicopter "if possible" and two experienced BLM hands, Jim Byers and Mike Hayes, whom he had seen on his previous fire. That message, timed at 4:05 A.M., would be waiting for Pete Blume when the fire-management officer arrived at work, but it would be evening before Byers and Hayes arrived at Storm King Mountain, too late to hike to the fire that day.

Blanco drove to the BLM office in Glenwood Springs and told the assembled hands to expect no helicopter, at least early in the day, which meant leaving camping gear behind and spending six daylight hours hiking up and down the mountain. It would be a long, tiring day, but one short on work hours. They headed out in trucks, a band of seven—Blanco and his partner, Brad Haugh, and Michelle Ryerson and her four hands—and minus Sam Schroeder's crew, not the thirty they had hoped to be.

They drove to the South Canyon exit of I-70, parked and

looked into the entrance of a brush-choked gully known locally as Hell's Gate, the source of the other place names. Hell's Gate had narrow, sheer sides, was choked with brush and seemed to lead nowhere.

The crew picked up their tools and started off. Blanco led the way, marking a path with flagging tape.

He led them into a labyrinth of twisting, eroded gullies known collectively as the East Canyon, a broad area between Hell's Gate and Hell's Gate Ridge. The crew followed dead-end washes, turned back and tried again. They would clamber onto a bench and then be confronted with a near-vertical drop-off. In places they had to pull themselves along on all fours. Their boots slipped in the loose shale; branches came out by the roots when they grabbed them for support. They climbed blind most of the time, catching only an occasional glimpse of the peak of Storm King.

At 8:19 A.M. Blanco's radio crackled, and Grand Junction's BLM dispatch office confirmed receiving his 4:00 A.M. request for help. They had no news of reinforcements.

Just below Hell's Gate Ridge the crew entered a patch of Gambel oak. The trees, ten to twelve feet high, had branches so numerous and interwoven that they resembled cobwebs. The canopy blotted out the sky.

The slope steepened this close to the ridgetop, and several crew members came to a dead halt. The fire had crept into leaf litter under the oak near them. Brad Haugh figured that a gust of wind at that moment could fan the flames, start the oak blazing and send fire sweeping over the crew. Haugh called to Ryerson to get her crew moving up and out of the oak, and she did.

Blanco's team broke onto the ridge about 9:30 A.M, more than three hours after they had started into the East Canyon, and took a much-needed break. Blanco walked along the ridgetop in search of a site for a helispot, a temporary landing zone in case a helicopter ever arrived. He picked out a flattish place several hundred yards back, or north from Hell's Gate Point, and set the crew to clearing it.

Haugh and Loren Paulson picked up the crew's two chain saws, which required the most skill to operate. The others cleared away branches and hacked the loose soil with shovels and Pulaskis, combination axes and hoes named for Edward Pulaski, a hero of the Big Blowup of 1910. In that year a firestorm blackened 3 million acres in Montana and Idaho, taking eighty-five lives and becoming the most famous fire in Western history.

The crew had been working for several hours when Haugh's chain saw threw a snap-ring holding the drive sprocket; he tried slipping it back on, but it kept popping off.

"Finally we said to hell with it," Haugh said.

The helispot had been cleared by then. Blanco told the crew to start a fire line off the west side of the ridge. The idea of a fire line is simple enough: clear a path a few inches to many feet wide around a fire, often right up against the flames, to keep it from spreading. In theory the fire dies out when there is no more fuel to burn; if the fire jumps over the line, the process starts again. Sawyers lead the way, swampers clear the brush, and diggers finish the job.

Blanco's plan alarmed Paulson, a nine-year Forest Service veteran. He told Blanco that they were heading into a situation warned against in fire manuals, namely, cutting line down a hill while fire burns below. Blanco agreed that it was worrisome, but he pointed out that sometimes there is no other way to fight a fire. Blanco told the crew to start the line and "keep their heads up."

Paulson led the way with the one good chain saw. He had gone about two hundred yards when something "went pinging off in the wilderness." The "ping," by a coincidence beyond odds, was the breaking of the same item that had snapped off on Haugh's saw, meaning that there could be no cannibalizing of parts.

Blanco called Grand Junction dispatch and this time asked for help of any kind. Moments later one of the emergency firefighters, Janie Jarrett, twisted her ankle, not a serious injury but one that cost the crew another hand. The next day, July 6, Jarrett ran errands with a truck, but even from that distance she would see enough of the South Canyon fire that by the end of July 6 she would determine "to quit firefighting due to this incident," as she later told fire investigators. After giving her statement, she drove off in the first automobile she could borrow, never to be seen again in Colorado.

In Grand Junction Pete Blume was having the same bad luck finding help as Butch Blanco was having fighting the fire. Blume had left a note for dispatchers the night before, July 4, saying he expected twenty-nine fresh smoke jumpers to be flown in to Walker Field by noon on the fifth, ready for assignment. Based on that, the BLM district dispatchers the morning of the fifth ordered ten of those jumpers to be sent immediately to the South Canyon fire.

By early afternoon no new jumpers had shown up at Walker Field. Blume decided at 2:20 P.M. to send an air tanker, and one started for the mountain. That plane, though, was diverted en route to another fire considered more pressing. It took more than two hours before a tanker, T127, made the run to Storm King.

There were plenty of big, troublesome fires on Blume's district on July 5, though overall there were two fewer than on July 2— thirty-six compared to thirty-eight. The Pyramid Rock fire, which had threatened ranches on the Fourth, had grown from 74 to 180 acres. Blume had declared Pyramid Rock the district's "first concern" for the fifth and put in a request for a special team to handle it along with about twenty small fires nearby, not including Storm King. The team arrived, leaving Blume free by nightfall on July 5 to declare a new top priority for the next day. It would be the South Canyon fire.

AS TANKER 127 came in sight of Storm King Mountain, the pilot, Randolph W. "Randy" Sullivan, wondered why he had been sent at all. The fire had spread, fingering down slopes and creeping along the ridge, becoming more a job for a helicopter with a water bucket than an air tanker. When he made a pass over the flames, a downdraft tugged at his plane, and he pulled up in a hurry. If he tried to fly lower, he figured, he could wind up with a wing stuck in the mountain.

Sullivan radioed Blanco to talk things over, but by then Blanco had new worries. Blanco asked Sullivan to see if the fire had burned to the foot of the western slope of Hell's Gate Ridge.

There was a gully at the bottom of the slope, a potential wind tunnel capable of magnifying any air current. If the fire had spread that far, there could be explosive consequences.

Everything looks okay, Sullivan told Blanco; the fire is still up near you. Sullivan told Blanco he was going to try to drop retardant on another slope, the one overlooking I-70. There was fire on that side of the ridge needing attention, and he would have more room to maneuver in the broad gorge of the Colorado River. Blanco agreed.

Sullivan began his run, aiming for flames about halfway down the slope. On the approach his plane hit an air pocket, or "sinker," and dropped like a stone. Sullivan slammed the throttle to full power and managed to pull out.

He circled for another try and this time found smooth air. The retardant fanned into a pink spume behind his plane, parallel to but a good fifteen yards below any flames.

Sullivan made one last run, attempting to edge T127 farther up the slope. The second half-load splattered on top of the first. Sullivan radioed Blanco and complained that the job really should be done by a helicopter, not a tanker. Blanco replied that he'd love to have one.

Sullivan turned for home. He radioed the Western Slope Coordination Center with his advice and heard a terse "Roger" from the dispatcher. That reply probably meant a zero chance for a helicopter, Sullivan concluded.

Asked later by investigators to sum up his mission, Sullivan replied, "The whole operation was what tankers call a 'cluster fuck.' "

AS SULLIVAN FLEW toward Grand Junction, Walker Field was being transformed from an overworked, understaffed outpost to a turnaround terminal for smoke jumpers and hotshots. The arriving jumper crews had only minutes to suit up, listen to a quick briefing and grab a frozen canteen before heading out.

The first extra jumpers arrived in late afternoon, a crew of eight from Missoula, Montana, where jumping has it deepest roots, from the beginnings of smoke-jumping—the airplane used on the first fire jump ever, July 10, 1940, took off from Missoula—to its one great tragedy, the Mann Gulch fire. On August 5, 1949, in a remote gulch in west-central Montana, a fire made the leap across the bottom of a gully and exploded in terrain and under circumstances hauntingly similar to unfolding events on Storm King Mountain. A crew of fifteen smoke jumpers and a wilderness guard who had been a jumper raced the fire to the head of the gulch; only three survived.

Before Mann Gulch the smoke jumpers had a perfect record of no fatalities from flames, and after Mann Gulch the record was the same: From 1949 until the summer of 1994, for four and a half decades, no smoke jumper was killed by fire, not one, though three died in jumping accidents. Mann Gulch had become a marker with a message: This must never happen again.

The Missoula jumpers spread their gear on the sparse lawn in front of the ready shack in Grand Junction and began suiting up. Rick Blanton, the jumper coordinator at Walker Field, told them that the drought had pushed burning conditions "several weeks ahead of where they should be," but said nothing about winds. It would be another hour before Chris Cuoco issued his evening forecast with a wind warning, which Blanton did pass on to jumpers arriving later. At this hour, 5:00 P.M., Blanton gave the crew the assignment he had been holding since morning: Storm King. Do a good job for Butch Blanco, Blanton said; he uses jumpers a lot.

The Missoula crew was eager for work. They had just spent two weeks on a detail in Santa Fe, New Mexico, without fighting a single fire. Plenty of fires had burned around them, but in open, sagebrush country, easy for fire engines to reach. The jumpers, after a flurry of attention from local media, had killed the time reading, working out and lying in the sun. Things were so slow, they were told on the Fourth of July that they were being sent home to Missoula; they learned when they boarded a plane the next day that they were headed for western Colorado instead.

Their natural leader was Don Mackey, a man known as the "heart and soul" of smoke-jumping. Mackey was big for a jumper, six feet tall and 185 pounds, with muscular legs built for the mountains. His face was round, boyish and handsome.

His fellow jumpers counted on Mackey as a hard worker, ex-

ceptionally skilled with a chain saw and able to teach others. He
had become one of the instructors for rookie training in Missoula.
But even more, they counted on Mackey for a lift, for his ability
to smile, start spinning a story and turn a nasty fire or a rainy day
in the woods into a memorable occasion.

Mackey was raised near Missoula at the mouth of Blodgett
Canyon, a gaping chasm in the Bitterroot Mountains, a region
so harsh that when Lewis and Clark crossed it in 1805 heading
west, they ate their horses to keep from starving. By the twentieth
century, elk and bear had been driven out of the valleys and into
the mountains, but the Bitterroots remained isolated, a place
where folks loved their country but felt no similar affection for
its government.

Don's father, Bob, had moved the family from Sacramento,
California, to Montana when his son was eight. Don and Bob
shared a love of the wilderness, hiking and hunting the mountains
behind their home, more brothers than father and son. By the
time Mackey was a teenager, he was running his own trapline
and putting together a collection of old-time traps that would
one day become a side business. Don had a single fear, that it all
would end and the family would move back to a city.

As Mackey zipped into his jumpsuit, he could see the family
brand where he had inked it once on the inside of his collar and
once inside his jump pants, a black capital M with the numeral
six attached to one leg. Nothing mattered more to him than
family. The brand stood for the six Mackeys: Don; his parents,
Bob and Nadine; and his three younger sisters, Susan Lorraine,
Jan Carleen and Mary Ann, "all beautiful, all with two children
and all in good shape," as Don once described them.

He and his mother were close, but Nadine encouraged Don
to follow his father. Mackey grew up convinced that he had a
responsible role to play around women and would be adored for
doing it.

This belief had been shaken by a marriage that had ended in
divorce the previous November. He'd tried to do the right thing
by Rene, persisting in the marriage for six years as he and Rene
became more distant, but things still hadn't worked out. They
wound up blaming the breakup on strains caused by his absence

all summer, every summer—jumpers joke that their divorce rate is 100 percent, if you count multiple divorces. The experience left Mackey bewildered that his best efforts with a woman he loved could have ended in failure.

At the age of thirty-four Don Mackey had a reputation in the Bitterroot Valley and beyond as a hunter, a teller of tall tales and a firefighter. After eight years with the smoke jumpers he had won a coveted career appointment guaranteeing six months' employment a year, a living in Montana's hardscrabble economy.

"He was one of those people born out of their time," said John A. Bird, his father's partner in the construction business. "He should have been born a hundred years ago, but he wasn't. He didn't complain about it; he made the best of it."

One night shortly before heading to New Mexico, Mackey had taken his two children, Leslianne, who would be six on July 7, and Robert Stewart, four, named for Bob Mackey, over to his parents' house. They spent the night in a child's cabin Bob had built in a patch of cottonwoods next to a creek. Don and Bob stayed up into the small hours trading tales around a campfire.

Don needed money for child support and was hoping for a change of luck from the rainy, jumpless season of the year before. There were worse places to start the season than New Mexico, Don said.

"I'm sure glad I'm not going to Colorado," he told Bob and Nadine before he took off. "That's where people get hurt."

Though the crew in New Mexico came together by chance—Mackey was there because had missed another assignment due to a foul-up in orders—they all knew each other, some well. Kevin Erickson, like Mackey a Bitterrooter, had been Don's best friend and six weeks earlier had become his brother-in-law. Don had introduced Kevin to his sister, Jan Carleen, after she and her first husband divorced, and the couple now lived in a house built by Bob Mackey on the family acreage.

It was Don who had convinced Kevin to become a smoke jumper, preparing Erickson for rookie training by running and digging practice fire line with him. "Don was more like a brother to me than my own brother," Erickson said.

Sarah Doehring, the lone woman on the crew, had been

friends with Don since he'd been her instructor in rookie training. Doehring had a long face, squarish chin and brown hair sheared off at shoulder length. Her hands were strong, red and rough. She had gray eyes, alert and appealing. Doehring had come west from upstate New York, looking for a place that would welcome a no-frills woman. In the off-season she planted trees under government contract, replacing what fire destroyed.

Even those rooted in the old ways, when the woods belonged to men alone, spoke admiringly of Doehring as "not very big but tougher than a twenty-penny spike."

Jumping drew together a wide assortment of talents and backgrounds and made them one. Keith Woods, one of the few blacks in smoke-jumping, had been a tuba player in the band when he was recruited from Mississippi Valley State University at Ittabena, Mississippi. The Forest Service had sent a team to the campus in 1988 as part of a program to increase minority representation in fire work. Woods's parents had warned him against the Northwest region with its Northern European culture.

"It was their worst nightmare," he said.

But Woods was excited by the idea of traveling and fighting fire. He harbored no social grudges and found himself accepted, at least by the jumpers. When he completed training in 1991, someone had to tell him that he had made history as only the second black ever to become a Missoula smoke jumper.

The crew had its own helicopter expert, Sabiño "Sunny" Archuleta, who had been on fires with Rich Tyler, foreman of the Western Slope helicopter. And it had its own legal counsel, Quentin Rhoades, heading toward his last year of law school at the University of Montana in Missoula. It had a black sheep, Sonny Soto, who was fired from jumping a year later for getting into a fistfight on duty, and a boy-next-door, Eric Shelton, who was planning to leave jumping to give marriage and a full-time job a chance.

The crew zipped into padded jumpsuits made of Kevlar, the same material as bulletproof vests. The jackets had high Elizabethan collars to protect their necks. The baggy pants closed with zippers running from ankle to waist. A separate strap protected the crotch.

They each slipped into a parachute harness secured around the chest and legs, and bent at right angles while a fellow jumper attached the compact parachute to the harness, checking and re-checking the heavy clips.

They snapped on a reserve chute in front, and below that a pack or PG bag containing personal gear—snacks, a hard hat, gloves, extra clothes, even a book for campfire reading. Each picked up a helmet shaped like a football player's but with metal screening across the face. They stuffed an assortment of other gear into gaping leg pouches—a 150-foot letdown line, water bottles, a two-way radio—and waddled to the plane swinging their legs as though in plaster casts.

They found seats aboard a DHC-6 Twin Otter airplane in reverse jump order, Doehring and then Mackey taking the last seats because they were numbers two and one on the jump list. Doehring had missed Blanton's briefing, instead making a quick run to the unisex washroom. Jumpers fill up on water before a fire to offset the coming drain on body fluids. But answering nature's call in a jump plane means a lot of unsnapping and no privacy regardless of gender.

The plane rolled down the runway at 5:20 P.M. It rose until through its tiny windows and open door the jumpers could see the smoke of many fires spreading a haze of pearl gray and dusty gold over the mountains, as lustrous as the embers of a fading campfire. Inside the plane it was hot, noisy and uncomfortable. The engine drum and the whoosh of wind made conversation a chore.

The jumpers sat in sweaty isolation inside their suits. Mackey, who always wanted to know what was going on, found a flight helmet with a built-in radio and listened to the air traffic. This was his world, not a never-to-be-recovered dream of the West but a living place where he and others went out like explorers to fight fire on a mountain. It was freedom bought with high technology, but it was still freedom.

"I'm glad he did it, stayed with being a smoke jumper," said his father, Bob, who had tried to join the smoke jumpers when Don did. "I always told him, 'If you find something you really believe in, stick with it no matter what.'"

As number one on the list, Mackey was smoke jumper in charge, just as Steele had been on the Wake fire. It was his first fire jump of the '94 season, as it was for the others—in fact, their first in almost two years, due to the wet '93 season. It was Mackey's first fire jump since being promoted.

On the radio Mackey listened as the spotter, Sean Cross, talked to Randy Sullivan in T127 flying back from Storm King. The terrain was so steep, Sullivan complained to Cross, that his retardant drops had done no good. He warned Cross that he had seen no place clear enough to drop jumpers.

Mackey hollered to the others that the fire was in steep terrain.

"Steep? We want it steep! Give us steep terrain!" the shouts came back. The hot current of an adrenaline rush pumped through their veins.

Worse than that, Mackey reported, there were power lines nearby.

"We want power lines! Give us power lines!" they shouted.

Cross, the spotter, thought Sullivan's report sounded bad enough that they should consider landing at the Glenwood Springs Municipal Airport and busing the jumpers to the mountain. But that would mean ground-pounding, the jumper elite hiking like ordinary mortals to their fire.

Ten minutes away from the fire Cross raised Blanco on the radio. Blanco said that his crew was having problems, and a load of smoke jumpers would be more than welcome. Blanco had been a parachutist himself, serving with the 82d Airborne during the Dominican Republic incursion in 1965.

The plane banked, and for the first time the smoke jumpers saw the South Canyon fire. The sight quieted them. The fire had grown to more than thirty acres, substantial by any measure; wisps of smoke rose in a spider-web pattern from burned-over portions, while flames ringed its edges.

Doehring, who had kept silent during the shouting, was startled to see an interstate highway directly below. Why hadn't the fire been attacked earlier? she wondered.

From the air, the scene was a kaleidoscope of fire, earth and sky sorting itself into a different mosaic for each of the jumpers. Some remembered seeing a "good deal," a sure thing for overtime but small enough to bring under control overnight; others thought it suspiciously large for eight jumpers.

The unmistakable shape of inevitable disaster had yet to form on the mountain side. Its seeds now lay scattered in indistinct forms and patterns, some not even physical, of a building cold front to the north, a dulled sense of emergency among Colorado fire managers, a tendency for key messages to go unheeded or unheard and the pride of a forty-five-year-old record of no smoke jumper lost to flames.

Mackey and the air crew thought Sullivan correct; there was no obvious jump spot. The spotter, Cross, had the pilot fly once, twice and finally a half dozen times over the site. They found one possibility, a small opening in the oak brush at the dip where Hell's Gate Ridge joins the mountain peak, the same spot on which Allen Bell had focused his camera the night before from Canyon Creek Estates. They agreed to try it.

Cross radioed the news to Blanco and went to the rear of the plane. Streamers showed erratic but light winds, calm enough for jumping. Cross, a careful man, had the pilot fly one last pass to make sure everyone was talking about the same spot. It was so small that Cross offered Mackey and Doehring the chance to jump "single stick," meaning one jumper per pass instead of two, though that would double the effort for him and the pilot. They said no thanks, they liked being jump partners.

Mackey put his foot on the step in the door, felt the slap from Cross and stepped into nothingness. He started counting by thousands, and by thousand-three he felt the static line yank the cover off his parachute. By thousand-four he had a canopy over his head. Engine noise had numbed his ears, but now he heard small sounds again: the whoosh of air past his face, the flap of his chute, the receding drone of the aircraft engines. The cushion of the sky replaced the jar of the airplane.

Sarah Doehring followed Mackey by seconds, and they immediately shouted at each other to make sure they wouldn't bump. Mackey outweighed Doehring by fifty pounds, and they yelled more out of habit and comradeship than a real fear that they would collide. After making sure of Doehring, Mackey followed the practice of a lifetime and fixed his eye on the target.

When Mackey was a teenager, he, his father and John Bird, his father's partner, had lived out a crazy dream for two middle-aged men and a boy—they had answered a newspaper advertisement to go skydiving. The ad directed them to an airport in the Bitterroot Valley, where they received a few hours' instruction before going aloft. Bird went first. An old Cessna single-engine airplane had been modified to accommodate jumpers, the rear seat and door taken out and a wheel fender removed.

To jump, Bird backed out the door and stood on the wheel while the pilot braked to keep the wheel from spinning. It was a clumsy maneuver at best, and Bird, a tall, beefy man, had a difficult time with it. At three thousand feet, with a big white X on the ground for a target, Bird let go.

Arms and legs flailing under the chute, Bird looked to those on the ground like a water beetle in distress. He sailed straight for an airport hanger that rose before him like a canyon wall. Bird tugged madly at the chute lines and at the last moment swooped away, tumbling to the ground in a nearby alfalfa field.

Don Mackey was beside him in a moment.

"How was it?" the boy asked breathlessly.

"Goddamn it, I *never* want to do that again," Bird said in exasperation, and he never did.

Don went up next. On his first-ever jump, he went out in businesslike fashion, feet together and eyes on the target as instructed. He hit the X on the nose, the only one that day who did.

FOUR

As SARAH DOEHRING floated above Storm King Mountain, she watched the canopy of Mackey's parachute angle toward the jump spot. Flappy winds caught her own parachute and pushed her toward I-70. This is a first, she thought, jumping in sight of a four-lane highway. She pulled her steering lines and maneuvered back on course as Mackey's parachute crumpled to the ground below her.

On this, the 155th jump of his career, Mackey landed on target, the only jumper this day or the next to do so. Doehring landed close enough to Mackey so they could have exchanged yells, but he was already on the radio with the jump plane, reporting their arrival and the tricky winds.

The other jumpers floated in. A wind gust caught Quentin Rhoades's parachute, and he emerged with a bruised hip and wounded pride. Kevin Erickson's parachute snagged in a tree. He clucked to himself about poor chute-handling and made a silent vow to do better next time.

When Keith Woods crashed into oak brush, it triggered a rush of unwanted memories for him—of a fire that moved like a bounding animal, flames that exploded and men who ran with nowhere to go. In the summer of 1990 he had been on an engine crew sent to a fire in a state park in Utah.

The blaze had started as a neglected campfire. By the time Woods's crew started off, it had spread to 150 acres; by the time they arrived, an hour and a half later, it had scorched 5,000 acres. It burned in Gambel oak, the same thing he had landed in on Storm King; the oak is almost impossible to ignite most of the time but burns like fury when dry. The blaze leapfrogged as winds of up to sixty miles an hour tossed embers ahead of the flames. One giant leap fatally trapped a bulldozer operator and a deputy sheriff. On Storm King Mountain in the middle of the oak patch Woods said out loud, to no one in particular, "I sure hope this fire doesn't make a run."

It was evening, about 7:00 P.M., before all the crew had landed. With dark approaching, they quickly piled their gear, leaving some parachutes where they had tangled in the oak.

Mackey radioed Blanco. The BLM foreman by then was half-way down the mountain with his battered crew and their useless chain saws. There would be no face-to-face meeting between the two, no detailed briefing, no talk of plans for the next day. Mackey and his jumpers, although used to operating alone, now found themselves in a kind of limbo. The fire belonged to them for the night, but Blanco retained overall responsibility with the title of incident commander. Mackey's title of smoke jumper in charge suddenly carried a lot of undefined responsibility. His fire-fighters needed a plan, more personnel and equipment and a weather forecast specific to Storm King Mountain, but it was unclear at that moment who was responsible for providing those things.

"I wasn't worried," Blanco said later. "He's [Mackey's] a jumper, he had a load of jumpers on it, very knowledgeable people with a lot of experience." For Blanco that explained everything. The smoke jumpers were the best; they came "self-contained" with food, camping gear and chain saws, and if they needed anything else, they would ask and it would be airlifted to them.

The smoke jumpers picked up their tools and set off along the top of Hell's Gate Ridge looking for the helispot, the temporary landing zone for the helicopter that Blanco had picked several hundred yards north of Hell's Gate Point. They hiked half a mile, clambered up a rock outcropping and walked onto the helispot, an exposed point with a good view of the western slope of Hell's Gate Ridge, I-70 and the Colorado River as it flowed toward the Grand Hogback.

From the helispot they could see the fire spread in fits and starts, downhill and sideways. Smoldering flames covered the western slope of the ridge directly below the helispot, including where Blanco's crew had started their fire line.

Mackey thumbed his radio.

"West line overrun," Mackey reported to the BLM's Grand Junction dispatch office. "West side active."

WITH DARK APPROACHING, the smoke jumpers looked forward to cooler temperatures, increased humidity and no winds. They would switch on headlamps in an hour or so and work through the night. They made a snappish remark or two about how little work Blanco's crew appeared to have done—the helispot looked to some as though fire, not firefighters, had cleared most of it—but their general mood was upbeat. They would have a line around the fire by midnight; after that the job would be a "piece of cake."

Avoiding the smoldering western slope, the smoke jumpers started down the opposite slope in the direction of Glenwood Springs, into the East Canyon. They planned to work around the ridge in a clockwise direction until they were above I-70, and from there make a fresh approach to the western slope to circle the fire.

They ran into trouble from the first step. The East Canyon was far steeper than it had appeared from the air. Their boots slipped; rocks and logs rolled down around them.

Storm King Mountain includes in its varied geologic makeup an extension of the Maroon Formation, well known and much feared by Colorado mountain climbers. More than 250 million years ago, in the time of the Ancestral Rocky Mountains, an enormous apron of sediment, reddish with iron ore, washed off the Uncompahgria Range to the south and extended as far as

Storm King Mountain. The sediment remains dangerously loose, or "poorly cemented" in geologic terms. The Maroon Bells, the famous twin peaks near Aspen named for the formation, have claimed many lives, earning the nickname "Deadly Bells."

Adding to the instability, the rock that makes up Hell's Gate Ridge tilts down toward the Colorado River, ready to slide, a result of the upthrust of the modern Rocky Mountains about 65 million years ago—the same tilt on the opposite side of the river secures the rock against the mountains there. A severe mud slide took place as recently as 1977 in Glenwood Springs on the Storm King Mountain side of the river gorge.

The terrain slowed the smoke jumpers enough that they needed headlamps by the time they started cutting a fire line. Cones of light swung back and forth as the crew moved forward in procession. Glowing logs cartwheeled down the slope, sending up showers of sparks when they struck the ground.

Keith Woods and Kevin Erickson, working next to each other, heard a heavy thudding and jumped back. They felt a wind as a boulder flew between them, and they locked eyes in alarm.

The crew crawled through a series of rockslides that ran at right angles to the slope, perfect chimneys if the fire flared up below them. The angle of the slope grew steeper, the ground more treacherous until they came to the lip of one rockslide too broad and deep to attempt in darkness.

"There's nothing on this hill worth getting killed over," Mackey said, repeating the remark several times. No one needed coaxing to turn back.

They retreated the way they had come, the route straight up being too steep. When at last they stepped onto the helispot again, a cool breeze met them. Several peeled off T-shirts drenched with sweat and placed them to dry over smoldering stump holes.

They pushed foil-wrapped meals into embers and tuned their radios to the weather broadcast of the National Oceanic and Atmospheric Administration or NOAA, regional broadcasts intended for a general public, which never include the Red Flag Warnings for firefighters. A cold front was pushing out of

Montana and Idaho and would be in Colorado the next day, July 6.

Cold front on the way? Hell, it's already arrived, several thought. After balmy nights in Santa Fe, the mountaintop felt unnaturally chilly.

A more specific forecast for the area around Storm King Mountain, with an ominous update, never reached the jumpers. In Grand Junction Chris Cuoco's evening forecast announced that the cold front was moving faster than expected: "The cold front will arrive earlier, about eight hours sooner than expected. . . . Red Flag Warning." There had been ten Red Flag Warnings for the Grand Junction District since June 1, including every day since July 2, but all the previous ones were for lightning. For the first time Cuoco now warned of strong winds.

Cuoco predicted that the front would reach northwestern Colorado at 9:00 A.M. the next day. It would move through Grand Junction between 2:00 and 3:00 P.M. and arrive at Storm King Mountain about 5:00 P.M., preceded by gusting winds. That forecast was faxed to BLM dispatch offices at 7:20 P.M., the same time the smoke jumpers were starting down the eastern side of the ridge, but was never broadcast by the Grand Junction BLM.

From the helispot the crew could see the fire stretching down the western slopes in long fingers, surprisingly active for that time of night. Beyond to the west were the lights of the town of Rifle. To the east Glenwood Springs was hidden behind ridges, but the night sky reflected the urban glow that joined starlight in a luminous haze. Far below, the crew could hear the rap of a truck's diesel engine from I-70 and the rattle of a coal train from the Denver and Rio Grande Western railway tracks across the Colorado River, an odd collection of sights and sounds for backwoods firefighters.

July 5 had been so dry that thunderstorms failed to develop for the first time in four days, and the humidity remained low. The temperature, which had reached 91 degrees at 5:00 P.M., when the jumpers were flying to the mountain, fell steadily to an overnight low of 58 degrees, registered at a weather station in Rifle.

The humidity dropped from 22 percent at noon to a bone-dry 11 percent at midnight. By 4:00 A.M. it had risen to 20 percent and by 7:00 A.M. to 31 percent, short of a normal recovery, before heading down again.

Mackey radioed the BLM in Grand Junction at 11:00 P.M. to make an unhappy report. Blanco's fire line didn't exist. He and his crew had been forced to retreat. The fire had grown to fifty acres, large for smoke jumpers, and was continuing to spread. They would need a lot more people by early morning to have any hope of bringing it under control the next day. He needed the best crews available, hotshots if possible, because of the rugged terrain. He asked for two crews, or forty people.

The night dispatcher, Thelma "Tuckie" LaDou, noted Mackey's request at 11:10 P.M.: "Mackey—Need two Type I crews." LaDou said that the district would do what it could, but she was not a supervisor and made no promises.

The jumpers prepared to bed down for the night. They had no shelters, and the chill made Sarah Doehring wish she had brought more clothes. She found a smoldering tree on the edge of the helispot and curled up next to it for warmth. Keith Woods did the same thing next to her.

At the foot of the mountain in Canyon Creek Estates, Allen Bell once again took his video camera into his backyard. He trained it on the slope over I-70, where fingers of flame extended nearly to the highway.

"A little fire last night, but look at it now," Bell narrated. "It was almost out this afternoon."

The flames shot up.

"That's taking off again!" Bell said. "That lower half could burn this way." If the fire spread into the gulch west of Hell's Gate Ridge, Bell said, then "there's one big mountain over there that's gonna be—" It was the same danger he had noted the night before, and this time he did not bother to finish the sentence.

THE COLD WOKE Quentin Rhoades about 2:00 A.M. He stumbled back to the jump spot and retrieved a sweater from his gear. When he got back to the helispot, Mackey was sitting up watching the fire, and Rhoades joined him.

A tree below them popped like a firework as its sap overheated. Mackey remarked how active the fire was for nighttime, and how much bigger than when they had first seen it. They talked for a minute or two about what they would face in the morning, agreeing that it would be a serious battle. Rhoades went to catch what sleep he could.

Mackey remained awake. He didn't need the title smoke jumper in charge to keep a watch for others; for better or worse, after growing up as the older brother of three younger sisters, keeping watch and holding people together had become one of his roles in life. There was a payoff at home and around campfires: a ring of admiring faces when he told stories.

One time he had lost a jump partner, Roger Archibald, on a small fire in the Bitterroot Mountains not far from his home—or more accurately, Archibald had become lost, having gone looking for his jump gear after nightfall. This particular mountain was made up of finger ridges with sheer sides. After a time Archibald, fearing he would fall off one, halted and began calling for Mackey.

Mackey was too far away to hear, but, concerned about his partner's long absence, he took his headlamp to the edge of a finger ridge and waved it back and forth. Archibald saw Mackey's signal "like a lighthouse beam across a sea of mountains" and followed it back to their campfire.

In an epilogue, when the Forest Service later raised questions about Archibald becoming lost, Mackey made a formal affidavit that explained away the incident as a common one on fires; even elite firefighters need a keeper once in a while.

BLANCO AND HIS crew were bone tired by the time they pulled into the BLM's West Glenwood office around dark the night of July 5. It had been a frustrating day, short on sleep, long on hiking and with little to show at the end. Blanco once again picked up the telephone and made a round of pleading telephone calls. He reached Cheney at the BLM's Grand Junction District dispatch office and told him that the arrival of the smoke jumpers had improved general prospects, but he needed yet another crew and a helicopter in the morning.

"He felt comfortable with it," Cheney said. "That's why we weren't in a panic mode, because he wasn't."

Cheney's "we," the BLM's fire managers, had plenty of other worries, such as the Wake fire, which continued to burn out of control. Governor Romer remained at the scene and had become involved in an unpleasant media confrontation with George Steele, though neither party had sought it.

Romer, with television cameras and reporters in attendance, came upon Steele as he and his crew tried to stop the Wake fire from overrunning an old coal mine. If flames reached the coal, it could burn for decades—a coal-mine fire on the Grand Hogback, near a place that had come to be called Burning Mountain, had been smoldering in 1994 for almost a century, the occasional wisp of smoke visible from Storm King.

As the governor and his entourage approached, Steele heard Romer telling a reporter that the fire was under control. The reporter turned to Steele and asked if that was correct. Steele pointed to a column of smoke billowing from the approaching fire.

"This fire's not contained or controlled," Steele said grimly, and the contradicting quote made the evening news. It was not to be the last confrontation involving the two men.

The BLM's Grand Junction District had no direct responsibility for the Wake fire, which was in the BLM's Montrose District, though it drew away shared resources. With a special management team now handling the most troublesome fire on the Grand Junction District, the Pyramid Rock fire, Blume late on July 5 was free to nominate the South Canyon fire as his number-one priority for the next day, Wednesday, July 6.

At 8:25 P.M. on the fifth, Blume made a formal request to the Western Slope Coordination Center for two additional crews for the South Canyon fire to be ready at dawn. Blume asked for Type II firefighters, less capable than smoke jumpers or hotshots.

This time the order went through to Boise. Less than a half hour later, at 8:50 P.M., NIFC assigned two Snake River Valley crews, migrant agricultural workers based in Vale, Oregon, who would have to be called together and bused to Boise Airport for a flight to Colorado, a process sure to take many hours.

At 9:15 P.M., Blume ordered a helicopter for the South Canyon fire for the next morning. It took only eight minutes to assign Rich Tyler's unit from the Western Slope center, helicopter 93R, or 93 Romeo as it was called on the radio.

When Mackey reported in at 11:00 P.M. requesting Type I crews, someone paid attention. At 11:21 P.M. an order for two Type I crews went out from the Western Slope center, canceling the order for the Type II Snake River Valley crews. With the demands for fire crews growing, only one Type I crew was immediately assigned to the South Canyon fire. The log entry reads "Prineville Hot Shots, w/bus."

FIVE

THE PRINEVILLE HOT SHOTS awakened at 6:00 A.M. on Wednesday, July 6, on the high-school football field at Fruita, a farming center eleven miles from Grand Junction. Around them sprawled other hotshot crews with fun names like Zig Zag and place names like Redmond, shuttled out from Walker Field to establish the beginnings of a fire city. The Prineville crew had spent many nights together on the playing fields of recessed schools.

Unlike smoke jumpers, the twenty members of a hotshot crew stay together all season, often year after year. They work, eat, sleep, shower and take time off as a unit—and when trouble strikes, they stick together. There were five women on the Prineville crew, adding to a sense of family. Seven of the males lived in a pair of apartments next door to each other, the Love Shack and the Puppy Love Shack, in Prineville, a desert town twenty-five miles northwest of the exact geographic center of Oregon. Among the males were two brothers, Tony and Rob Johnson, and Levi Brinkley, one of triplets; Joel, Levi's brother, was later to become a Prineville Hot Shot.

In their way hotshots are as elite as smoke jumpers, capable of battling fire anywhere at any time, but unlike jumpers they specialize in big fires requiring teamwork. Jumpers in a bragging mood say they'll never go to hell because they'd put the fire out; hotshots respond that hellfire is more their size. The term "hotshots" made its formal appearance in 1951, when the first groups were put together as an alternative to recruiting in saloons, and by 1994 there were sixty-six crews nationwide consisting of more than thirteen hundred hotshots.

The Prineville Hot Shots, formed in 1979, had a reputation for being hardworking and fun. Most were in their mid-twenties, a few years younger than the average smoke jumper. They shared a restless spirit, a desire to taste danger under controlled conditions—recklessness with rules and always in company with others. Many had taken off a few years between high school and college to serve in the military or bum around. Nearly all now worked to pay tuition costs for college or graduate school; fire work has become a chief source of tuition money for the West's striving middle class. As they chanted the ten basic fire orders in training, the first order—"Fight fire aggressively, provide for safety first"— became transformed into "Fight fire aggressively, provide for *overtime* first."

On the Fourth of July they had been released with a "job well done" from a fire in southern Oregon. They camped in the woods for fun that night, thinking they were on their way home to Prineville, just as the Missoula jumpers thought on the Fourth that they were heading home. The next morning over breakfast the hotshots learned they would instead be going to western Colorado. They greeted the news with joy: Prineville Hot Shots had fought fire in every Western state except Colorado and Kansas, so this meant another state pin for their hats, another checkoff for the crew record book. And it was Colorado! Home of Vail, the Denver Broncos, ski instructors and snow bunnies.

Tom Shepard, the group's superintendent, had a graying beard and customarily wore the full Prineville uniform: blue sweatshirt and cap, both embroidered with a coyote dancing over a flame, and green fire pants. Shepard's quiet, reassuring manner drew crew members back season after season and even inspired an occasional Father's Day card.

The foreman, Bryan Scholz, wore a bristly mustache and hair cropped short. Scholz drove the crew with determination and a paternal affection he tried to conceal behind a bluff manner. His superiors found him prickly and difficult.

Scott Alan Blecha, six feet two inches tall and weighing more than two hundred pounds, set the pace for the crew, chattering nonstop as he worked. Blecha carried a homemade root ripper, a short shovel with a blade welded at a ninety-degree angle, too heavy for almost anyone else. After four years in the Marines, Blecha had graduated cum laude in engineering from the Oregon Institute of Technology and was earning money for an advanced engineering degree.

Jon R. Kelso had been a Missoula smoke jumper before returning to Prineville for family reasons and joining the hometown hot-shot crew. He was everybody's kid brother, earning the nickname "Tadpole," but with his extensive fire experience he had been made a squad boss. Kelso had a degree in wildlife science and had completed his first year in civil engineering at Oregon State University, earning straight A's.

Douglas Dunbar, honor student, all-star baseball player and award-winning saxophonist from McKenzie Bridge, Oregon, was in his second season as a hotshot. A month earlier, his mother, Sandy, had told him of having an anxiety dream about his job: She had been driving through a dark forest when she came upon a "crummy," the local name for a crew bus. As firefighters descended from the bus, one came over and peered into her windshield. His blackened face frightened her awake.

This was the rookie fire season for two crew members; on a hotshot crew newcomers can be sandwiched between experienced personnel. By contrast, smoke jumpers, with their emphasis on individual initiative, must have previous firefighting experience. Kathi Walsleben Beck was a free spirit who had fallen head over heels in love with firefighting. She grinned so hard her cheeks dimpled. Her auburn hair, which flowed over her shoulders as fiery as autumn leaves, was barely contained by a scarf of Indian-blanket design, its only concession to uniformity being the background color, Prineville blue. A senior at the University of Oregon earning a psychology degree, Beck dreamed of starting her own wilderness-adventure group.

The other rookie, twenty-eight-year-old Terri Hagen, had completed airborne training while an Army Reserve medic, though her mother, Jill, thought her still "naive and vulnerable." She was a member of the Onondaga tribe of the Iroquois Nation, raised on a reservation in New York State until the age of nine. Hagen had made her mark on the crew her first day digging line. When a male veteran asked her, with excessive politeness, if she could work a little faster, Hagen straightened and replied, "I was in the military; you don't have to kiss my ass. If I'm not doing something right, let me know and I'll fix it."

Hagen was there to earn tuition money to finish a degree in entomology at Oregon State University. "She would run *after* things we would run *away* from," said one crew member.

By 1994 women had been formally recognized members of fire crews for more than two decades. The first all-women crews were put together in Alaska in 1971; acceptance came more slowly for women in the lower forty-eight states. Once in those early days two women from Nevada's Silver State Hot Shots were washing off fire dirt in the restroom of a lodge near Las Vegas. A group of elderly women came in, and one, taken aback by the sight, remarked, "You mean they let girls fight fires?"

One of the hotshots, Kathy May, looked out of blackface and replied, "No, ma'am, we just make sandwiches and sweep rocks."

A decade later, in 1981, Deanne Shulman became the first woman smoke jumper after a struggle with both sexism and the requirement that recruits weigh at least 130 pounds, a standard later reduced to 120 pounds to accommodate women. When news of Shulman's achievement reached jumpers in Alaska, legend has it, one veteran left the mess hall and threw up.

The hotshots, with their family culture, have accommodated women more easily than have the smoke jumpers, but even so, women hotshots have to be tough, sometimes more so than the men. An early incident involving such a woman is still remembered by both sexes. Hotshot Jody Crawford worked in the off-season guiding and packing for elk hunters in Wyoming. One evening Crawford came into the lodge wearing a work shirt and tight jeans, a round can of chewing tobacco visible in outline in one hip pocket and a half-pint of whiskey in the other. A young man staggered out of the lounge, patted her on the whiskey bottle and remarked, "I sure would like to get me some of that!"

She knocked him flat with one punch, looked down and said, "If you learned to ask nicely, hon, I might think about it."

By 1994 it was not unusual for women to account for a quarter of a hotshot crew, as they did on the Prineville Hot Shots. Tamara "Tami" Bickett, beautiful enough to have been a Strawberry Festival princess in high school, had been introduced to the hotshots by Kelso; they had fallen into puppy love and become engaged, but caught themselves in time and became best friends instead. Bickett, like Kelso, was a squad boss.

Bonnie Jean Holtby, the daughter of a smoke jumper and granddaughter of a firefighter, was the winner of a string of scholar-athlete awards and a woman of deep religious conviction. Kim Valentine, married to a firefighter, was shy and careful, a graduate of George Fox Christian College.

The women stirred romantic fantasies for the men in their first days on the crew, but after weeks of shared sweat, bathroom jokes and morning breath, they came to be more like sisters than dreamland lovers.

"After you've all been out and gotten dirty and tired, she's your sister, and anybody who messes with her is in trouble," Blecha told Michael Thoele for his well-regarded book, *Fire Line: Summer Battles of the West*. "That's not always something the women appreciate. They don't always want someone looking out for them."

Physical testing has presented problems. Women have worked to increase their upper-body strength, about 50 percent of a man's on average before muscular conditioning, but still lag by a wide margin. The standard for physical fitness in 1994 for federal firefighters was a step test using a bench, set a few inches lower for women to account for physical differences. The subject would step up and down for five minutes at a given pace, and the subject's heart rate would be measured at the beginning and end. A passing grade was the same for both sexes, the equivalent of finishing a one-and-a-half-mile run in eleven minutes and forty-five seconds or less. The run could be substituted for the step test.

Smoke jumpers have far more demanding requirements. All jumpers must run one and a half miles in eleven minutes or less and perform seven pull-ups, twenty-five push-ups and forty-five sit-ups. The killer is the pack test, three miles on flat ground in ninety minutes or less while carrying a 110-pound pack, after finishing the run and calisthenics. Those standards replaced more extreme requirements at some bases, in particular in Alaska, after women joined jumping.

More recently, hotshot supervisors, concerned that the step test measured only aerobic fitness and not muscular power, requested a substitute that better reflected the work of a fire line, hiking with a heavy pack and digging with a hand tool. In response, Brian Sharkey, who developed the step test during the 1970s while teaching physiology at the University of Montana, came up with a replacement. Today, federal firefighters, regardless of gender, must be able to hike three miles on flat ground carrying a forty-five-pound pack in forty-five minutes or less.

"The old hotshot image was Rambo with a shovel," said Scholz, the Prineville foreman. "Firefighters have died living up to that image. The presence of women has changed that environment to one of cooperation. There is no specific fire-suppression task which requires a penis."

It took the day of July 5 for the hotshots to find their way by bus and airplane to Fruita. When they arrived, they were assigned a fire for the next day many miles to the south near Durango.

On Wednesday morning, July 6, they spent several hours waiting for a bus, and when one finally showed up, it brought a collective sigh of appreciation. It was a luxury model, complete with toilets and air-conditioning, unheard-of comforts for a fire crew. At the last minute, the Prineville Hot Shots learned that another unit would be taking their assignment and the bus.

At 8:00 A.M. an aging yellow school bus lumbered into sight. The driver cranked open the door and hollered, "PRUNEY-ville!" The Prineville Hot Shots climbed aboard with the laughter of their fellow hotshot crews ringing in their ears.

They had a new assignment: Storm King Mountain.

✶   ✶   ✶

MACKEY'S CREW ON Hell's Gate Ridge welcomed the first rays
of sun on July 6 after a poor night's sleep. They had done little
but lie around in Santa Fe, and one night on the ground was no
hardship. As they stumbled to their feet, they already had the
beginnings of a fire look, with sooty clothes, ash- and sweat-
streaked faces and bloodshot eyes. They placed meals to warm
and water to boil over smoldering stumps. Keith Woods made a
breakfast of hot cocoa and freeze-dried spaghetti with *sauce pi-
quante*—firefighters learn to tolerate the taste of char, or carry a
bottle of hot sauce, or both.

The fire had spread during the night in a random, spider-web
pattern along the western and southern slopes of the ridge. The
fire was less active in the East Canyon, where the crew had been
stopped the night before, a relief because no one had the slightest
desire to attempt that side again. Flames descended the slopes in
lazy dollops and made short upslope runs—an unusual amount of
fire activity, the jumpers remarked, for the cool of morning.

The jumpers, watching the fire as they ate, agreed that they
would need more help than they previously had thought.

Mackey radioed the BLM's Grand Junction District and found
the same dispatcher, Tuckie LaDou, on duty. He told LaDou
that the fire had been "active all night," and he was going to
need a fixed-wing aircraft, or "eye in the sky," to monitor its
progress. He also asked for a heavy-type helicopter to haul gear
and personnel and make water drops. Helicopters have been the
versatile workhorses of the fire world ever since 1949; the first
recorded use of a helicopter for fire duty came in Mann Gulch
the day after the fire there, handling transport duties.

It was not yet 5:30 A.M., and LaDou was virtually alone in the
office. Flint Cheney, the lead dispatcher, had left the office
around midnight and would return by 8:00 A.M., but Mackey had
a fire to fight and needed an immediate answer.

LaDou telephoned the Western Slope Coordination Center
and reached her counterpart, Donna Olshove, the night dispatcher. LaDou then apparently conducted what amounted to a
three-way conversation among herself, Olshove and Mackey.

Olshove wanted to know if Mackey thought that a single helicopter could perform observation duties and other tasks. She later told fire investigators she had not wanted to "mix fixed-wing with helicopter," apparently fearful of a collision. Fire managers, though, routinely assign aircraft different elevations over a fire to keep them apart, a practice Olshove either ignored or did not know about. Under normal circumstances supervisors would have made such a decision, but the dawn hours during a busy fire season are abnormal by definition.

Mackey agreed to the compromise, Olshove later told fire investigators. Mackey's call was logged at the BLM's Grand Junction District at 5:28 A.M.: "Helicopter w/long line for gear removal. Fire grew during the night. Wants air observer."

Two minutes later the log of the Western Slope center notes that the request for an aerial observer was "canceled per Donna," and the light helicopter already assigned, 93 Romeo, was substituted for both the heavy helicopter and aerial observer.

Mackey needed a plan to begin the day's work without over-committing his crew before others arrived. He decided to cut a fire line along the half mile of ridgetop back toward the jump spot. This would provide a holding line if the fire made a run up the western slope, and give the crew a chance to pull their gear together to be hauled out by 93 Romeo. No sight is so welcome to smoke jumpers as a helicopter with a cargo sling saving them a bruising pack-out.

Nobody offered a better idea, and the jumpers went to work.

BUTCH BLANCO DROVE to the BLM's West Glenwood office at first light, rousted his crew and told them to pack camping gear for several nights. They would remain on the mountain this time and not repeat the previous day's round-trip fiasco.

Michelle Ryerson, the flight attendant–turned–engine foreman, had trouble getting up after another night of sleep disturbed by the ringing telephone. The callers kept asking, Why is the BLM taking so long with a fire we can see out our picture windows? Ryerson had no answer.

Before starting, Blanco called the BLM's Grand Junction District for a weather update. At this early hour he received a forecast from the previous evening, the latest one available, which carried a Red Flag Warning, as it had four days in a row. This time the forecast also contained more welcome news, a hint of rain in elevations above 7,500 feet. With Storm King peaking at 8,793 feet, Blanco told everyone to be sure to pack rain ponchos.

Blanco headed by truck for Hell's Gate for the second day with eleven people, including himself and his partner, Brad Haugh, having left behind the emergency firefighter, Jarrett, nursing her twisted ankle. He had gained five more BLM locals, including the two, Mike Hayes and Jim Byers, he specifically had requested, and another man who had been with them, Derek Brixey.

Blanco cautioned them to take it easy hiking to save their energy for the fire. They passed through Hell's Gate and started up the East Canyon about the time Mackey finished his exchange with the dispatch office, at 5:30 A.M.

Blanco, who scouted ahead, arrived first on Hell's Gate Ridge and immediately checked in by radio with the BLM dispatch office in Grand Junction. For once he heard good news: He could keep the eight smoke jumpers already on the ridgetop, and a second Type I crew had been found, another eight jumpers who should be parachuting in shortly. By the time Blanco finished the conversation, it was 9:15 A.M.

Blanco's crew tuned their radios to the public NOAA channel. The regional forecast called for pleasant weather, windy and cooler, with temperatures rising only to the mid-eighties. It would be sunny in the morning and partly cloudy in the afternoon. Winds would be fifteen to twenty-five miles an hour with some stronger gusts, diminishing toward evening. Temperatures would drop to possible record lows of fifty to fifty-five during the night as the cold front took effect.

Blanco went to find Mackey, and for the first time the IC and smoke jumper in charge met face to face. Mackey and the jumpers already had cut the beginnings of a fire line along the ridgetop and begun to gather their gear. The two supervisors quickly agreed they needed to conduct an air reconnaissance before finalizing an overall plan. "There was hardly any way to see that whole fire at one place," Blanco told fire investigators later. "You couldn't see down the sides of it; there's just no way you could."

As Blanco and Mackey talked, they heard the *buzz-whop* of a

helicopter and a small green-and-white craft came into view—
93 Romeo. The helicopter seated only six but had a 737-
horsepower engine with plenty of lift. Another half hour had
slipped by; it was 9:45 A.M.

Upon arriving over Storm King, 93 Romeo's pilot, Dick
Good, a Vietnam War veteran, made a quick turn over the fire
so he and the helicopter foreman, Rich Tyler, could get their
bearings. They then flew to the meadow at Canyon Creek Estates, where the helicopter's ground crew, who had driven out
in trucks from Grand Junction, were setting up a helibase, a landing zone and maintenance area.

Within minutes a second aircraft, J17, hove into sight from the
direction of Grand Junction, bringing the second load of jumpers.
For the first time since the fire had started nearly four days before,
the tide of events began to run in favor of those trying to put it
out. The time was 10:23 A.M.

Mackey's crew stopped work to perform the ritual of watching
fellow jumpers float in. The new arrivals were a mixed group
pulled together for Storm King, hailing from a variety of jumper
bases and wearing different-colored hats and packs. There were
two Missoula jumpers, Billy Thomas and Tony Petrilli, the latter
being one of Don Mackey's best friends; Petrilli and Mackey were
roommates when both found themselves at the Aerial Fire Depot
in Missoula. There were two jumpers, longtime partners, from
the North Cascades base in Washington State, Eric Hipke and
Dale Longanecker. There were four jumpers from the McCall
base in Idaho: Jim Thrash, forty-four, a bearded professional guide
and packer in the off-season, and his partner, Roger Roth, a
member of the Oneida tribe of the Iroquois Nation who worked
in the off-season at a federal refuge for panthers in Florida; and
a last set of partners who shared the same first name, Michael
Cooper and Michael Feliciano. The parachutes landed scattered,
as they had the day before, a testament to the mountain's tricks.

Blanco and Mackey boarded 93 Romeo at the ridgetop to
make their reconnaissance flight. The pilot, Good, directed
Blanco and Mackey into seats on the left side of the craft and
then flew wide counterclockwise circles to give them the best
view of the scene.

Below them lay the western drainage, a deep chasm between Hell's Gate Ridge and the low ridge immediately to the west in the direction of Canyon Creek Estates. The western drainage was nearly two miles long and shaped like a giant funnel, widest near the peak of Storm King and tapering to a small opening just above the Colorado River.

The vegetation was thickest on the upper sides of the drainage near the ridgetops. Stands of oak and PJ gave way to grass, brush and scattered trees farther down the sides. Water had carved the bottom of the drainage into a deep V, filled with dead brush and trees, which became more pronounced closer to the river.

At the time the supervisors went aloft, the fire covered more than a hundred acres on Hell's Gate Ridge, mostly on its western side. The doubling in size since early morning was not a good sign. It descended in a simple, spreading pattern from the original ignition place, Hell's Gate Point. Small flames crackled and light smoke rose in random patterns, then subsided. The fire stewed, or "skunked around," as 93 Romeo's pilot, Good, said later, "little burned spots and unburned spots all over, little pockets of fire everywhere."

The helicopter's engine clatter made discussion difficult, and after a few circles Good turned the craft for the meadow at Canyon Creek Estates. There, Blanco, Mackey and Tyler got out and found a spot away from the helicopter to talk things over.

They now had a fair-sized but not overwhelming crew, twenty-seven counting Mackey's eight smoke jumpers, the eight jumpers from the second load and Blanco's eleven locals; six more if you added Tyler's team running the helicopter. The twenty Prineville Hot Shots, scheduled to arrive hours earlier, should be showing up any minute.

The problem the supervisors faced was how to fight the fire beginning from the top of the ridge. Having everyone hike or take the helicopter down the mountain and start from below, at the bottom of the western drainage, ran so contrary to their actual situation that they gave it scant, if any, attention, though after the fire many touted this as a safer and more sensible alternative. On July 6, after a strenuous effort to put twenty-seven people on top of Hell's Gate Ridge, none of the supervisors argued for starting over again from a different place.

Mackey outlined two possibilities, both of which involved starting a fire line at the ridgetop. The first called for working downhill on the west flank of Hell's Gate Ridge as close to the ragged edge of the fire as possible. This was the same strategy Blanco had tried without success with a much smaller crew the day before. Mackey, before going aloft, had marked a starting point for a west-flank fire line in a slight saddle about 250 yards north of where Blanco's original line had been, to take account of the fire's subsequent spread north.

The slope dropped off precipitously for a hundred yards from that point before reaching more level ground. Mackey's proposed west-flank line would proceed down the first steep hundred yards and then another hundred yards along less steep terrain until it reached below the lowest edge of the fire. It then could be turned at a right angle toward I-70. From the turn it would have to cross a series of spur ridges running at right angles off Hell's Gate Ridge to encircle, or "hook," the fire.

This maneuver, a line cut right up against a fire, sounds excessively dangerous but often is preferred as the safest available tactic. Crew members always know firsthand what a fire is doing and can keep "one foot in the black," meaning they can step immediately into a previously burned or black area for safety. That particular technique became famous in the wake of the Mann Gulch fire, which included the most spectacular, and most saddening, example of "one foot in the black" in history.

With flames roaring up Mann Gulch toward the smoke jumpers, the foreman, R. Wagner "Wag" Dodge, lighted a separate fire in the path of the main one. Dodge stepped into its ashes, waving and yelling at his crew to join him, but they thought that the foreman had gone "nuts" and passed him by. The main fire caught all but the two youngest, Bob Sallee and Walter Rumsey. They clambered straight up and out of the gulch, using Dodge's fire as a buffer against the advancing flames. Dodge lay down in the ashes of his own fire and survived, but never again could he bring himself to jump from an airplane.

Mackey's second possibility for Storm King Mountain was to attack the fire indirectly, starting downhill from the top of Hell's Gate Ridge but keeping the fire line well back from the flames. While this sounds safer, it seldom is. Creating a swath of unburned fuel gives fire a chance to gather itself for a run. The usual technique when employing this option is to imitate Wag Dodge in Mann Gulch and, using fusees the way Dodge used his matches, burn out the green swath in sections as the line advances. This technique takes extra time and many firefighters, both of which were in short supply on Storm King by midmorning on July 6.

Mackey, Blanco and Tyler agreed that they had enough people for a direct attack on the fire, but probably not for an indirect one. Blanco, who had the final say as incident commander, thought a direct attack, downhill and up against the flames, to be "prudent." Tyler confined his opinions to what the helicopter crew would be doing, but he was an experienced fire hand and had a say.

"We flew the helicopter to recon the fire, and we picked line locations and everything like that," Blanco later told fire investigators. "It was a cooperative." The investigator asked, Who had the "ultimate responsibility"? Blanco responded, "Would have been mine."

It was decided, then: They would launch a downhill, direct attack on the west flank of the ridge, keeping one foot in the black as much as possible. And they would carry plenty of fusees. Once they had hooked a line around the fire, they had another option: a onetime massive burnout of the dried and partially burned fuel remaining where the fire had fingered down. They could help the fire kill itself.

Mackey and Blanco flew by helicopter back to the ridgetop. Along the way Mackey radioed the jumpers and told them they might as well start cutting line from the point he had flagged.

The jumpers found his mark and looked down. Below them lay an unbroken mass of oak. Because the fire was advancing in fingers, the first stage of the west-flank line would be indirect, with no black to step into; that would come later, once the line swung under the fire and headed toward the Colorado River.

Until then they would be in a sea of green oak punctuated by an occasional pine tree.

They radioed to Mackey, "You sure you want us to do that? Go down that side?"

Mackey told them that the route looked better from the air.

"Are there any safe spots down there?" they asked.

Mackey said that the vegetation opened up toward the bottom of the gully, creating a safety zone. "It's pretty sparse; it doesn't look too bad."

The jumpers had no way to see what Mackey was talking about. Their view of the bottom of the western drainage was partially blocked by a spur ridge that ran perpendicular to Hell's Gate Ridge and then curled below them.

Jumpers expect to have a say in critical matters; in fact, they insist on expressing opinions on every issue, critical or otherwise. They radioed Mackey, "We're going to wait for you to come down here and explain some stuff to us." It was not an outright revolt, but it was stiff resistance.

Kevin Erickson, who held the temporary rank of squad leader and thus, technically, outranked Mackey, took a few steps off the ridgetop to scout around. By the time Mackey returned, Erickson was filled with questions.

The unease was general, but the other jumpers kept quiet as Erickson and Mackey walked off a short way to talk. Some figured it was a "brother-in-law thing," others simply held back. Later this moment would be remembered with collective self-reproach as a time when smoke jumpers denied their basic nature, the urge to leap into battle, whether it be an argument, a barroom brawl or a raging fire, holding back nothing.

Is there any safe zone if things go bad? Erickson asked. Is it really safe in the bottom of the drainage?

Mackey acknowledged that it was "pretty brushy" near the ridge where they were, but again said that the vegetation became "sparse" toward the bottom of the gulch. If the fire made a run, he reasoned, it would race uphill and away from the bottom of the drainage, and thus away from the firefighters.

"Let me have a big crew and we'll do this, we'll do fine," Mackey told Erickson.

The issue hung in the air. Finally Mackey said, Why not let the fire have a say? "Look at what it's doing now," he said.

Flames sputtered in leaf litter, rising to a height of no more than three or four inches. Smoke was light and wispy. The oak leaves shone a reassuring green. Winds were light.

Erickson admitted that the fire "didn't look that bad"; the next thing he knew, he had agreed to the west-flank line, and everyone had sensed it without hearing the conversation. After days of delay, critical commitments now came in split seconds. Longanecker, smoke jumper in charge for the second load of jumpers and a veteran of nearly five hundred jumps, led the procession as smoke jumpers began to step off the top of Hell's Gate Ridge.

Sarah Doehring wondered with black humor why the jumpers, and not Blanco's crew, were heading downhill. "Hey, can't we flip for this?" Doehring said. When Mackey asked the jumpers for volunteers to remain on the ridgetop and help with cargo, Doehring stepped forward. She joined Sunny Archuleta, the jumpers' helicopter expert, who had been requested as a cargo handler by Rich Tyler.

Blanco asked Ryerson if she would volunteer to reinforce the jumpers and take her engine crew down the west-flank line. She flatly refused.

"I'm too tired, and I think my crew's too tired to go down there," Ryerson said. Instead, Blanco's entire crew of eleven went to work widening the line started earlier that morning along the ridgetop. They also began clearing a second helicopter landing site, designated H2, closer to the jump spot and better protected from winds than the first helispot, now designated H1.

Those jobs would occupy them the rest of the day.

While Mackey and Erickson had been talking, a log on the slope nearby had caught fire. The jumpers and BLM crew had quickly cut a line isolating the spot, and it posed no further threat. But one of the jumpers, Quentin Rhoades, had taken it as a bad omen and privately resolved "not to go down the hill digging line."

Rhoades reconsidered, though, as he watched the other jumpers head off the ridgetop. I sure as hell am not going to stand around watching other people work, he told himself. Turning to Erickson, he jokingly asked if he could borrow a red cape.

"Super Sawyer!" Rhoades shouted and he, too, stepped off the top of Hell's Gate Ridge.

SIX

As THE YELLOW school bus carrying the Prineville Hot Shots made its way to Storm King Mountain the morning of July 6, a layer of clouds shaped like lines of turrets—a formation called *altocumulus castellanus*—warned of an approaching storm. The officials at the BLM's Grand Junction District had said nothing about weather that morning when they had given the hotshot superintendent his assignment. They had seemed overwhelmed by the fire activity, Tom Shepard recalled later, and "the information exchange was minimal."

The hotshots made slow progress. Their school bus labored well below the seventy-five-mile-an-hour speed limit posted on most of I-70. It was 10:00 A.M. before they pulled off to refuel at a gas-food station at Rifle, seventy-five miles from Fruita, providing the first shopping opportunity for the hot shots since their arrival in Colorado. They combed the aisles for snacks, soft drinks and Colorado hatpins. A clerk, Lorna Chinery, remembered them as being in high spirits, laughing and wondering if they would be fed before they went to work.

It was 11:00 A.M. by the time they came in sight of the South Canyon fire. The upbeat mood vanished.

"What the hell is that?" someone called out. "You've got to be kidding me. That fire's dead!"

They saw no towering forests, no crystal streams, no snow-capped mountains—no Colorado! Storm King Mountain, with its blotches of PJ and jagged rockslides, reminded them of the desert around Prineville. The fire was a "puffy, twenty-five-acre nothing," its most remarkable feature being its proximity to an interstate highway. Shepard snapped a photograph.

Bryan Scholz, the foreman, felt a pinprick of apprehension. He had seen the same thing a few weeks before, a routine brushfire on a steep slope, and that time the fire had exploded. It had happened during the crew's first assignment of the season, to fires on the Sierra National Forest in Northern California. They had hiked into a box canyon and joined other crews assembling in a meadow. Ahead they saw a line of flames at the base of a steep slope. The crew started to flop down in the meadow, the temperature being over a hundred degrees, but Scholz called them to their feet.

"I told them what was going to happen," Scholz said. "The folks on the other crews were looking at me like I was some sort of knucklehead. And it happened. The fire made one huge run from bottom to top in a minute, probably a good half-mile square."

This is a drought year, Scholz told the crew. "Learn the lesson now, when we don't have to pay the price."

The bus passed the South Canyon fire and headed for the BLM office in West Glenwood. The crew needed to pick up shovels, fusees and gasoline for chain saws, items too bulky or dangerous to carry on airplanes. They found one person, a clerical worker, on duty.

"We didn't know you guys were coming," she said.

"We're the elite firefighting crew," Scholz said only half jokingly.

"Oh," she replied with an innocent look, "are you smoke jumpers?"

They told the woman they were reinforcements for the South Canyon fire. "Don't call us smoke jumpers," Scholz said. "When the fire gets too big for them, they call us in."

The woman seemed glad enough they had come, remarking, "Nobody else wanted to go up there." But she still refused to open the BLM's equipment lockers without permission, a situation confirming the complaint in the 1993 internal audit that the BLM office in Grand Junction "controlled all the shots" and isolated the local office in West Glenwood.

While telephone calls were initiated to obtain permission, several hotshots went rummaging in the shop behind the office. The clock ticked toward noon, and then past noon. Janie Jarrett with her twisted ankle showed up to act as guide. By then the scroungers had found everything they needed except gasoline, which remained under lock and key. They set off with Jarrett leading the way, purchased their own gasoline and started for Glenwood Springs in search of lunch.

Jarrett's two-way radio squawked. Butch Blanco had tired of delays and wanted the hotshots on the fire immediately.

The school bus turned for the mountain.

There had been little work for Good and 93 Romeo after dropping off Mackey, Blanco and Tyler on Hell's Gate Ridge. Good wound up giving helicopter tours to children who appeared from the houses around the meadow. The children climbed over both the machine and Good himself while mothers snapped photographs.

The more the children played, the more uncomfortable grew Robert E. "Rob" Browning, Jr., a helicopter crewman with a gentle Deep South accent, on temporary assignment from the Forest Service's Savannah River Forest Station in South Carolina. Browning liked children but thought they had no place around helicopters. Steve Little, his partner from the Savannah River station, had drawn the assignment for the landing zone on the ridgetop, helping Tyler, and was waiting to fly up. Browning had been assigned to the helibase in the meadow. Browning asked Little if they could switch.

Sure, Little said. "Thanks, bud, I owe you one," Browning told him and patted his friend on the shoulder.

The Prineville bus pulled up at the rail fence circling the meadow. The hotshots formed a human chain and began unloading their gear.

Once again after long delays, key decisions were made in finger-snap time. The helicopter crew sorted the hotshots into groups of five for transport to the ridge aboard 93 Romeo. The hotshots had been broken up so often that they had modified their alphabetical manifest to spread out the leaders: Shepard was always first on the list, and there was a supervisor—either Shepard, Scholz or at least a squad leader or boss, the equivalent of a corporal—with every five or six hotshots. This time it backfired.

By chance the first five names on the manifest included most of the crew's heavyweights. Shepard, weighing 233 pounds including tools and personal cargo, was switched from the number-one position to the second load to even things up. He was replaced by Tami Bickett, who joined Kelso as a squad boss on the load. This took care of weight concerns. But it meant that neither Shepard nor Scholz, the true decision makers for Prineville, would be with the first group of five.

It was just after 1:00 P.M. when 93 Romeo took off for the ridge with the first five hotshots, minus Shepard. Blanco met them and told them they would be joining the smoke jumpers on the lower, west-flank line. When Shepard arrived with the second load minutes later, Blanco told him, in effect, I've talked with your outfit, and they're ready to go down the hill.

If there had been a moment to think, Shepard said later, he would have realized that his crew was heading into trouble. But in the final rush to the ridgetop, Shepard had not even taken time for an air reconnaissance. All Shepard could do was watch as nine Prineville Hot Shots picked up their tools and marched off the ridge, drawing reassurance from the knowledge that smoke jumpers already were down the west-flank line.

"There was nothing in Butch's tone to indicate urgency," said Shepard, who remained on the ridgetop in his customary role of lookout and coordinator. "It's just that it happened so fast."

As the helicopter started back for a third load, Don Mackey called on the radio. The fire had spotted across his line, he said, and he needed 93 Romeo to make a water drop. Good interrupted the shuttle of the hotshots and flew to the meadow to retrieve the water bucket and its cable.

Bryan Scholz and the rest of the Prineville crew stood idle in the meadow, watching the helicopter commence water drops. They could see the top two thirds of Hell's Gate Ridge, but the lower portion was hidden from them by the intervening low ridge. They couldn't make out the west-flank line at that distance, but the water drops indicated its general location. They heard a crescendo of growls as the chain saws of their fellow hotshots joined those of the smoke jumpers. Scholz looked at his watch. It was 1:30 P.M., and he and his half of the crew had become stranded listening to other people work.

"Those guys up on the hill were having a good time cutting stuff, getting bucket drops, and we were sitting on our keisters," Scholz said later.

Scholz impatiently thumbed his radio and called Shepard. He told him that his crew would arrive on the ridgetop a lot faster by hiking than by waiting for the helicopter.

Shepard by then had stationed himself on a knoll about three hundred yards north of the starting point for the west-flank line. From there he had a sweeping view of most of Hell's Gate Ridge. But the aspect farthest south, nearest the river, was blocked by a large spur ridge that began at the original helispot on Hell's Gate Ridge, H1, and ran down into the western drainage. Shepard told Scholz that the hike would be a lot longer than it looked, and he should wait for the helicopter.

Scholz scanned the mountain from Canyon Creek Estates for the exact location of the roaring saws. He saw instead a column of smoke, small but distinct, rising from the western drainage. The low ridge blocked his view of its source, but it seemed to be coming from near where the Prineville crew should be working. If the smoke column signaled that fire had slipped below the crew, real danger threatened. Every firefighter knows the rule: Never let a fire get below you on a mountain. Only bears and fires, not firefighters, can run uphill faster than down.

Scholz radioed Shepard a second time and asked if he could see where the smoke column was coming from. "Is anybody dealing with it?"

Shepard replied that he couldn't see that far, his view was blocked by spur ridges, but the jumpers should be right about there. The jumpers were on it.

A reporter, David Frey of *The Glenwood Post*, had arrived at the meadow to cover the event, and Scholz started to give him his lecture about fire behavior, pointing out the smoke column. "If the fire gets down far enough—"

A hot shot cut in. "Then we're looking at overtime!"

On the mountain, the smoke column traced upward, a mystery about to solve itself.

CHRIS CUOCO WENT out on a limb the morning of July 6 by issuing his forecast with a Red Flag Warning while other forecasters around the state issued lesser Red Flag Watches. There had been nine thousand lightning strikes in western Colorado over the previous several days, and the cold front would arrive earlier and with less moisture than expected. That was enough for Cuoco. He faxed his warning and forecast to radio dispatchers at 7:30 A.M.

The front should pass through Grand Junction between 10:00 A.M. and noon, Cuoco predicted, and should reach the area around Glenwood Springs by 4:00 P.M. It would be preceded by strong southwest winds, shifting to the northwest as the front passed. Gusts would reach thirty miles an hour.

By midmorning Cuoco had begun to have conflicting doubts, that his forecast might be ignored and that it might be wrong. He skipped breakfast. He posted copies of the forecast on a bulletin board at the Western Slope Coordination Center and kept popping out of his windowless cubbyhole to check the sky.

The American flag on the pole outside the center fluttered in gentle southerly breezes. Cuoco was peering anxiously out a window when Mike Lowry came by and mentioned that he had just seen Cuoco's forecast.

"You know, Mike, the cold front might arrive closer to noon than ten A.M.," Cuoco said, backpedaling.

"Don't start second-guessing yourself, Chris," Lowry admonished in a friendly way.

A half minute later the flag on the pole went slack. There was a moment of dead calm, and then it fluttered up again, now held aloft by northwest winds.

"There it is!" Cuoco said. "The front made it!" Cuoco checked his watch; it was 9:50 A.M.

Lowry slapped him on the back. "See?" Lowry said. "And you doubted!"

Cuoco took his weather gauges outside. The winds were only about five to ten miles an hour. He went back to his cubbyhole and made a series of telephone calls. Conditions were similar elsewhere in the path of the front, lighter initial winds than expected. But the front itself was proving more formidable than predicted, bringing with it gusts to forty miles an hour. And in Grand Junction the temperature was heading toward 90 degrees, which meant there would be no initial cooling.

The realization came to Cuoco that the front would strike with almost no warning. Half the fires in western Colorado could explode, and if someone didn't warn firefighters, they could be caught in harm's way.

At minutes past noon the main front provided the final confirmation, sweeping into Grand Junction with sudden winds just as the lightning storm had four days earlier. For Cuoco this was "an adrenaline-pushing, severe weather condition," the same as a tornado warning. There was no time to prepare and fax an update. He grabbed a telephone, scratching a note to himself on his now-outdated forecast: "Verbal update—12:30. No winds preceding. Go abruptly to WNW with front—no cooling."

He reached a dispatcher in the Montrose District who thanked him and said, "This is important stuff; we'll get it out to everyone." The warning was on the district's airwaves in plenty of time for firefighters to pull back or take measures to "windproof" their fires, including the Wake fire, which had grown to two thousand acres.

Cuoco telephoned the BLM's Grand Junction District and received an identical response, a grateful thank-you and a promise that the update would be broadcast on an urgent basis. But the district's outdated radio equipment, a "box with a knob" as dispatchers later described it, was overloaded with traffic. The dispatchers themselves were worn from overwork. The message became lost. No word of Cuoco's Red Flag Warnings for wind, or the telephoned update, was ever mentioned on the radio network covering the 1.8 million acres of the BLM's Grand Junction District.

THE WEATHER FRONT advanced across the Book Cliffs and into the valley of the Colorado River, past Battlement Mesa and on toward the Grand Hogback. It arrived in Rifle at 2:27 P.M., swirling around the cozy home, picket fence and vegetable garden of Maxine Myster, who had been making daily reports for the National Weather Service for twenty-eight years. Gusts registered forty miles an hour.

At Canyon Creek Estates, twenty-five miles to the east, Scholz felt a restless stir in the air. He thumbed his radio and called Shepard. Was the wind up there acting "squirrely" the way it was down here? he asked. Shepard replied that yes, he felt the same thing.

The shifting breezes had little initial effect on the fire. Smoke rose in lazy swirls from Hell's Gate Ridge. By 3:00 P.M. the fire was calm enough for Good to switch from making water drops and resume ferrying Prineville Hot Shots to Hell's Gate Ridge.

Scholz, as hotshot foreman, rode up on the fourth and final crew shuttle. He checked his watch as he stepped onto the ridge, noting that it was 3:15 P.M. He walked over to the other supervisors, Shepard and Blanco.

The plan, they told him, called for Scholz's nine hotshots to join Blanco's crew in "slicking out" the ridgetop, a term invented for the occasion, which meant clearing virtually the entire backbone of the ridge.

Shepard, who had been on the ridgetop for more than two hours, felt little unease; the fire, he said later, appeared if anything "too innocent." He had become briefly alarmed when Jon Kelso, the ex-Missoula jumper on the lower line, responded to a question by saying that the fire line wasn't in black at all. Kelso quickly explained that the line threaded its way between fingers of burned and unburned vegetation, going in and out of good black, and assured Shepard that things were going well. "There was no sense of concern in Jon's voice," Shepard said later.

The crew on the mountain was at its moment of greatest strength, numbering forty-nine plus a helicopter and pilot. Yet it had a command structure suited for the first hours of attack on a small fire, not a time-consuming battle over hundreds of acres. In technical terms the South Canyon fire had advanced beyond the "initial attack" stage, defined as a small crew, a small fire and one day's work. The next stage, "extended attack," occurs when a wildfire eludes control for more than one "operational period," usually twenty-four hours, and is still less than a hundred acres in size, larger for rural areas. The South Canyon fire had eluded control for four days, been fought for two days, and had grown to 127 acres by early afternoon on July 6, surpassing these conditions by a wide margin.

An extended attack requires more firefighters, from dozens to hundreds, and more equipment. It involves a management team with separate chiefs, all under the incident commander, for planning and operations and, if necessary, for logistics and finance. Smoke jumpers, if they remain on the fire at all, play a lesser role, as Steele did when the Wake fire grew into an extended attack.

In a written directive the BLM's Grand Junction District had further spelled out how to recognize when a fire had crossed the gray line between initial and extended attack. An extended attack would require crews and equipment "beyond those normally available from the district," and would still be "experiencing perimeter growth after the first burning period." The South Canyon fire met both those conditions *before* July 6.

It was the responsibility of Blanco as incident commander or of the fire managers in Grand Junction, Pete Blume and Winslow Robertson, to upgrade the South Canyon fire. In the early afternoon of July 6, none did. Robertson later told fire investigators that the management system was working "just fine" at that time. He said that the BLM's Grand Junction District routinely put fifty firefighters on a fire with a similar command structure, one incident commander supported by crew chiefs.

Robertson, Blume's eyes and ears in the field, had made no attempt to revisit the fire since July 3, even though by July 6 it had become the district's number-one priority. As events unfolded on the sixth, Robertson remained in the BLM office in Grand Junction, as did Blume.

Asked later by fire investigators why neither he nor Blume considered upgrading the fire, Robertson said it was because of the number and the high-priority of the other fires.

Blanco remained in overall charge at the scene, but he left the west-flank line to Mackey and to the smoke jumper in charge from the second load of jumpers, Dale Longanecker. Blanco was criticized that day and afterward for never walking the west-flank line, but his decision can be at least partially justified as that of a commander remaining in an observation post while delegating responsibility to his chiefs.

Scholz, being the last supervisor to arrive on the ridge, had the freshest eyes. At least once he spoke up about his growing alarm and impatience. But he was a number two, a foreman, outranked by Shepard and Blanco. When they told him to put his crew to work on the ridge, he did it.

Scholz got another chance to express his anxieties a few moments later, when Mackey called by radio and said his crew badly needed drinking water. Before anyone could speak up, Scholz grabbed two five-gallon cubitaners, or "cubies," of water and stepped off the ridgetop; he was going to see the bottom of that smoke column for himself.

As he started down the slope, Scholz noted that the overcast had been swept away and that the sky glistened. It remained calm, though, in the thick oak on the protected slope.

Scholz hefted the cubies, eighty pounds together, and wondered "what in the hell" he had gotten himself into. The weight pulled him off balance, and he lurched down the slope. Wicked-looking stumps, or "stobs," left by the sawyers threatened to punch holes in the plastic cubies, and in him, too, if he fell.

About fifty yards down the slope he passed two BLM firefighters, Brad Haugh and a swamper, Derek Brixey, who had been sent down earlier to widen the line behind Mackey's crew. That was the farthest down the west-flank line anyone from Blanco's crew ever descended. Scholz forged past them with barely a word, a man on a mission.

The fire line dipped and then rose to the crest of the first spur ridge, the one blocking the view to the bottom of the western drainage from the top of Hell's Gate Ridge. Scholz hoped to find his answer there, to be able to see the source of the smoke column, but when he arrived, he found the view blocked by more spur ridges. He noted that the fire, creeping through leaf litter, had dried out, or "cooked," overhanging oak leaves, creating tinder rather than safe "black."

Scholz pressed on. On the far side of the spur ridge he came upon one of the smoke jumpers from the second load parachuting in that morning, Roger Roth, who was struggling with a smoldering log. Scholz set down the cubies of water with relief and went to help.

The log glowed from end to end. Scholz and Roth wrestled it perpendicular to the slope to keep it from rolling farther downhill. That job finished, Scholz said, "I'll see you," and started to pick up the cubies.

"That's okay," Roth said. "You brought them down this far, I'll take them the rest of the way."

"Are you sure?," Scholz replied.

"I'm going that way anyway," Roth said.

Scholz paused. The job of transporting drinking water had been accomplished. Shouldn't he go back to check on the nine hotshots he had brought to the ridgetop?

But the other nine hotshots were only a few hundred yards ahead of him on the west-flank line. Weren't they his responsibility as well? And he had yet to accomplish his private mission, discovering the source of the smoke column.

The thought came to Scholz that he could have it both ways, go back up the fire line and still accomplish everything. He could check on his crew on the ridgetop and then scout the smoke column by walking along the ridgetop to the south, in the direction of the river. He could look or hike down into the western drainage from there. It would take only a few more minutes, a few extra steps. Besides, shouldn't he heed his own instincts and get out?

Scholz turned back.

"I would have gone down there," Scholz said later, "if Roth

had said, 'You take one and I'll take one.' Heck, I was going that way anyway. If he'd said, 'See you later,' I'd have humped 'em both down there, no problem. I can carry water to thirsty folks.

"Maybe I got scared. I don't—Maybe my subconscious is smarter than I and said, 'Get the hell out of there!' Which is extremely irresponsible if you're a supervisor and you've got people down there."

The last hundred yards back to the ridgetop was a climb in itself, a pitch so steep that Scholz had to grab stobs and pull himself hand over hand. This would be a real bitch, he thought, if you were in a hurry. He again passed Haugh and Brixey and said words to the effect it was getting hot below and they should consider pulling out.

When he regained the ridgetop, a west wind slapped his back. He saw several of his crew in a bunch, excited and nervous. One tossed branches into the air, which the wind grabbed and swept into the East Canyon.

Shepard's voice crackled over the radio. Mackey wanted more help on the west-flank line, the Prineville superintendent said. Could Scholz send down some of his hotshots?

Scholz shouted a single word into the radio.

"No!" he said.

SEVEN

THE SMOKE JUMPERS, their hard hats bobbing in the sea of vegetation, made slow progress clearing the west-flank fire line. Their chain saws jammed on the springy oak. The growth was so dense, they had to cut openings in it to stack brush. Only a few yards off Hell's Gate Ridge Tony Petrilli's saw threw its looped chain. Petrilli and Mackey's friendship, made closer as each had struggled through a divorce the preceding year, included rough kidding about saw skills: Mackey never let Petrilli forget he had been his grader for the sawyer-qualification test.

Mackey came by and offered to help with the chain, but Petrilli stopped him with a look. Petrilli reset it himself, revved the saw until it screamed and then took a slow, deliberate swing at a twig in the fire line. It sailed into the air.

"Did I do that just right, Don?" he said. Mackey laughed and walked off.

The jumpers cut the line down the first precipitous hundred yards and then along the more level ground. Now below the flames, they could turn in the direction of the Colorado River and begin the hook around the lower edge of the fire. They faced the first spur ridge running perpendicular to Hell's Gate Ridge. As they started to cross it, fresh smoke rose ahead of them, from behind the spur.

"Let's get out of here. This isn't worth it," Mackey called out, echoing himself from the night before. Glad of any excuse to withdraw, the crew turned back, all except the man in the lead, Dale Longanecker, who was acting as line scout, deciding where the fire line should go next.

Longanecker was far enough ahead to see smoke rising from a single flaring pine tree. He radioed Mackey that there was no need to retreat, the helicopter could handle this with a couple of water drops. Mackey radioed for 93 Romeo, which had been ferrying the first half of the Prineville Hot Shots to the ridge. The helicopter retrieved its bucket from the meadow and hovered overhead within minutes.

For once the sight of a helicopter dousing flames was unwelcome to a crew of firefighters, several of whom managed to climb nearly to the top of Hell's Gate Ridge before Mackey called them back. Grumbling a bit, they once again filed down hill.

Their morale received a boost a few minutes later when nine Prineville Hot Shots, the first half of the unit ferried to the ridgetop, came down the line behind them. Jon Kelso's old Missoula buddies gave him a big round of hellos. The jumpers and hotshots combined saw teams and formed a staggered line a dozen feet wide. The sawyers, standing nearly upright, mowed back and forth, leaving stob points at a murderous forty-five-degree angle. The swampers cleared branches, and behind them the diggers grubbed out an eighteen-inch-wide cup trench, the centerline of a highway, to catch rolling embers.

Mackey ordered Quentin Rhoades to fell the doused pine tree so it wouldn't reignite and cause more trouble. Rhoades's worries about downhill fire line vanished in the excitement of toppling the tree, a good two feet in diameter. It swayed a moment, then fell to earth with a satisfying thud.

It had been almost two years since Rhoades had worked on a fire, and he began to feel wonderful as he picked up the old rhythms. He was with a top-notch crew; the pace had quickened with the addition of the hotshots; the fire line relentlessly chewed into the oak.

Rhoades's saw had dulled, and as he stopped to sharpen it, Mackey happened by and asked if he needed a hand. Rhoades liked Mackey, though they saw each other only during fire season; Mackey could make you feel like a best friend the second time you met him. But Rhoades considered himself a first-class sawyer, and Mackey's offer rankled, whatever its intent. Rhoades bent to his task, muttering to himself that a smoke jumper in charge should have more pressing duties than filing saw teeth.

The crew on the west-flank line was now a sizable but not overwhelming number, an even two dozen: Mackey and six of his smoke jumpers—Sarah Doehring had rejoined the crew, leaving Archuleta to finish the cargo work at the new helicopter-landing site, H2—plus the eight fresh jumpers, plus the nine hotshots.

The crew was working parallel to and several hundred yards below the top of Hell's Gate Ridge, hard up against the flame edge wherever possible. The sights of civilization lay below— I-70, the railway and the Colorado River, which was playing host by this time of day to a flotilla of kayaks and inflated rafts. The fire line, though, amounted to a tunnel open to the sky, an isolated world of limited horizons. The crew could see neither the place they had started nor their destination, could feel no wind— in fact, could barely make out the fire smoking above them.

The slope on which they worked appeared level from the air but was proving a roller coaster of spur ridges extending at right angles from the top of Hell's Gate Ridge. After crossing the first of these spurs, the crew saw another, smaller rise ahead, not big enough to be called a spur. They began to yearn for a view of the whole scene, the slope and fire.

But when they crested the low rise, they confronted the most imposing spur thus far, a long ridge with a reptilian backbone of loose rock that extended from the old helispot, H1, to the bottom of the western drainage. It was the same spur blocking Shepard's view from the top of Hell's Gate Ridge of the telltale smoke column; it came to have its own name—Lunch Spot Ridge. It would be a while, though, before the crew on the west-flank line reached it.

The sun poured down. The air hung windless on the slopes. Mackey and his crew took a quick break and a rationed drink from canteens filled that morning. Mackey kept calling Blanco to request more drinking water, not wanting to spare one of his crew to make the trip back to the ridgetop, but so far without success.

Mackey was talking with Tami Bickett, the former Strawberry Festival princess, when he saw his brother-in-law, Kevin Erickson, coming along the line and went to join him. The two men flopped down next to each other. Mackey, eyes twinkling, asked Erickson, "What do you think of that one?"

Erickson smiled: He was a newly married man, but it was impossible to miss the doe eyes and hints of rich brown hair beneath Bickett's hard hat.

Still joking, Mackey asked Erickson who he thought was the incident commander on the fire, himself or Blanco.

"I don't know," Erickson said, playing the straight man.

"Me neither," Mackey said. "Do you want the fire?"

"I'll take it if you want to give it to me," Erickson replied, and for a moment neither man was joking.

"Naw," said Mackey after a pause. "We'll see what happens later."

The conversation left Erickson confused. Was Mackey criticizing Blanco for remaining on the ridgetop, a general too far removed from the battle? Erickson felt irritated about that himself. An IC is expected to make himself known to everyone on his crew, but after almost twenty-four hours on the fire Erickson wouldn't have recognized Blanco if he had popped up in front of him.

Mackey was questioning his own performance, though, not just Blanco's, and that raised broader questions. Why would someone as self-confident as Mackey offer to turn over his fire to another smoke jumper? It had been unusual enough that Mackey twice had called for a retreat, saying the fire wasn't worth anyone's being injured. Clearly the nature of the fire unsettled him—it was days old when he arrived, nastier than expected and growing, the opposite of what a smoke-jumper fire should be, namely small and new.

Mackey had a plan for dealing with the fire: the downhill line that had been approved by Blanco and Tyler. But it had been challenged by his fellow jumpers, and most of all by the smoke jumper closest to him, Erickson.

Mackey had received less assistance than requested, and what did arrive was late. Even at this hour, about 2:00 P.M., half the Prineville Hot Shots promised for early morning remained in the meadow at Canyon Creek Estates. The firefighters' informal motto is "Ask for what you need, make do with what you get," but this had become an extreme example.

Mackey's responsibilities had outgrown his qualifications, through no fault of his own. He lacked the formal training to boss a small crew, let alone handle a complex operation like the South Canyon fire—a far more experienced smoke jumper in charge, George Steele, had been relieved as incident commander on the Wake fire the day before, without fuss or prejudice, when it grew too large for him. What was Mackey supposed to do, relieve himself? He nearly did that when he asked Erickson if he wanted to take over the fire.

Mackey was at ease with individuals and small groups, and with hand skills such as parachuting, sawing and digging, but not with the supervision of half a hundred people, a helicopter and crew and more on the way. Being a smoke jumper in charge was intended to give jumpers a taste of supervision, not install unqualified personnel as chiefs of major operations.

Mackey and everyone else knew that the situation was dangerous and growing more so. Smoke jumpers, though, are paid to attack risky assignments, not sit on mountaintops watching their fires crackle below them. His plan for the fire could work if they moved quickly, if they received more reinforcements right away, and—above all—if the wind held off.

The moment between Mackey and Erickson passed, and they rejoined the others.

The crew advanced the west-flank line until it reached a game trail leading to the crest of Lunch Spot Ridge. The vegetation began to change from Gambel oak to more open PJ and a few stately Ponderosa pine and Douglas fir trees.

As they went up the side of Lunch Spot Ridge, the crew had a much-improved view of the bottom of the western drainage, their escape route if things went bad. The sight was anything but reassuring. The bottom of the drainage was a brush-filled gully a good 250 yards away. The vegetation indeed did thin in places near the bottom, but that was best seen from the air; from their perspective it looked like an unbroken brush thicket.

The view upward, toward the top of Hell's Gate Ridge, held no cheer either. The junction of Lunch Spot and Hell's Gate ridges was invisible four hundred yards away up a steep, brush-covered slope; this was the same burned and cleared spot where Mackey and his crew had spent the previous night, the original helispot, H1. It would make an excellent alternative safety zone to the bottom of the western drainage, but neither the Prineville Hot Shots nor the smoke jumpers who had arrived that morning had seen it for themselves.

The west-flank fire line now extended seven hundred yards back from Lunch Spot Ridge, over the low rise, over the first spur ridge and then along the final steep segment to the junction with Hell's Gate Ridge. It provided an escape route, but a long one, with the last hundred yards the most demanding.

It was midafternoon, the low part of the day for a worker but the most explosive time for a fire. After hours of sun and heat, the three sides of the classic fire triangle—oxygen, fuel and high temperature—edge closer together. Fire scientists later estimated the chances of an ember's touching off flames on Storm King Mountain at that hour on that day at 90 to 100 percent, though making the prediction after the event significantly reduced the risk to forecasters.

There had been no break for food since breakfast, which meant dawn for most of the crew. They needed a long breather by the time they reached Lunch Spot Ridge and, with few spoken orders, took one, breaking down into home units. Rhoades, Petrilli and a few other Missoula jumpers found a juniper tree shaped like a candelabrum on the crest of the ridge and sat in a row in its shade. Longanecker and his partner from Washington State, Eric Hipke, sat on the ground a few feet away. The Prineville Hot Shots and the other jumpers found rest spots farther back along the fire line. They ate mostly energy bars, nuts or granola, nothing to make them logy in late afternoon.

ERICKSON AND DOEHRING found a spot together on the west-flank line short of Lunch Spot Ridge. As they settled down, the Prineville Hot Shot who had been a Missoula jumper, Jon Kelso, came walking by. Doehring had wanted to talk to Kelso, but there had been no time to do so until now.

Doehring and Kelso had been in the same rookie jumper class in Missoula three years earlier. They had shared a lot of laughs, and Kelso had given Doehring a T-shirt as a remembrance. By luck she had it in her pack and had planned to show it to him over dinner. As Kelso came along the fire line, Doehring thought, Why not now?

"Look what I've got, Jon!" she said, and pulled out the shirt, a much-faded, tie-dyed red. They shared another laugh before Kelso rejoined the other hotshots.

As Kelso left, Mackey came by and stopped for a moment to chat with his friends. A fresh wind rustled dead oak leaves at their feet. Doehring felt a stab of anxiety and asked Mackey what they should do if the wind fanned the fire to life.

Mackey told her to go straight down, into the bottom of the western drainage, assuring her that the vegetation thinned out closer to the bottom.

That made sense to Doehring. As long as flames stayed above them, they could wave the fire good-bye.

Doehring felt comfortable on a fire when Mackey was present; she had trusted him ever since an incident in rookie training when she had been suspended in jump harness from a high tower. She was supposed to unhook herself and slide down a line to the ground, simulating the common problem of snagging a parachute in a tree. Mackey had been the instructor on the tower.

The parachute harness had had the feel of a straitjacket, and Doehring had tensed. She knew the reputation of instructors for screaming at rookies, and had vowed to quit on the spot if that ever happened to her. She thought of herself as a tree planter first, a smoke jumper second.

Mackey took her through the procedure as though telling a story, step by step, his voice soothing and rhythmic. Doehring found the letdown line in her leg pocket. She tied it to the parachute riser lines, unsnapped the harness and slid to the ground, safely if slowly, and afterward never forgot who had made it possible.

The freshening wind bothered Mackey, his assurances to Doehring to the contrary notwithstanding. He thumbed the transmitter on his radio and reported, "The wind's getting a little girly down here."

"*Girly,* Don?" Doehring said, laughing.

Mackey quickly corrected himself. "The wind's getting *squirrely.*"

Women were seldom far from Mackey's mind, especially since his divorce had made him a free man again. Every one of the Mackey children had wound up divorced, the three girls and now Don. Bob Mackey said he never had an ounce of trouble with his offspring until they married. Perhaps the closeness of the Mackey clan was part of the problem; it surely had been for Don's wife, Rene. She had never gotten along with Don's parents, and had appeared to resent Don's attachment to his family. He was always dropping in at the bank where his sister Jan was a teller, or stopping by his parents' house.

The family might have been a problem for the Mackeys, too. The girls' divorces were tied to the Montana economy—the husbands wanted to move on to better jobs—but the Mackey clan was tough competition for any young couple who belonged to it and were trying to establish their own lives.

Since his divorce, though, Mackey had found a girl who admired his family ties. Melissa Wegner was, like him, a firefighter, a squad leader on the Bitterroot Hot Shots. Wegner was blond, athletic and spirited. Mackey told her, teasingly, that she looked like a sunflower in her yellow fire shirt and green pants.

Melissa glowed whenever Don talked about his family or asked her along when he had his son and daughter, Robert and Leslianne, for a weekend or listened with her to a favorite song, which always made him teary, "Love Without End, Amen," George Strait's country ballad about a father's enduring love for his son.

Before heading to Santa Fe, Mackey had written Melissa a letter, one of the few he ever wrote, recounting a dream in which he played the dual roles of a fire and a firefighter battling over an oak tree. The tree represented Melissa. In the role of the fire Mackey advanced to consume the oak. As the firefighter he struggled to protect it. He called in an air tanker for a retardant drop, but the first attempt missed. A second run did the job, and the dream ended with Mackey exhausted but eager to do battle for her again. "God, I get so much satisfaction out of this job," the letter concluded. "Love, Don."

When neither was away on fire work, they shared a trailer in Darby, a tiny, picturesque town at the head of the Bitterroot Valley. Their friends predicted marriage, but it was too early for Don. There were other women in other towns. It was a bitter moment for Melissa when she later found out about them, though she had a forgiving nature and kept on loving him.

Mackey got to his feet and told Erickson and Doehring he was off to check the rest of his crew. As they prepared to follow, he stopped for a moment, as though something bothered him. He told Erickson as an afterthought that he had better head back along the line to make sure no burning materials had crossed it. Later Erickson would recall this moment and wonder if Mackey had been able to sense, if only dimly, what lay ahead.

Doehring felt uneasy enough about the fire to wish she were leaving with Erickson, but stepped into line behind Mackey. Mackey halted for a second time. "Sarah, why don't you go back and help Kevin?" he said. Doehring obeyed with a feeling of relief, asking no questions; she lost sight of Mackey in seconds.

Erickson and Doehring had walked a short distance when they heard a clattering below them. It could be a scurrying animal, rolling stone or tumbling firebrand. They peered into a tangle of gray oak branches and oily leaves and waited for smoke to appear. None did, and they moved on.

*   *   *

THE PARTNERS FROM Washington State, Dale Longanecker and Eric Hipke, ate hurriedly and left Lunch Spot Ridge in opposite directions, Longanecker to mark the new course for the fire line and Hipke to check on the morning's work. They had been near enough to the Missoula jumpers to hear them recounting their early-season exploits. Rhoades had kidded the others about their attentions to an attractive young lady at a saloon in Santa Fe. In his telling, the jumpers had surrounded the woman like a pack of "sexual predators" until Rhoades had gallantly steered her to safety.

The "Zoolis" sounded full of themselves, Hipke thought, but typical for smoke jumpers. It always galled him, though, to hear Missoula jumpers take on airs about their "storied past," as though a colorful history made them superior to other jumpers. What most bothered him, however, was the talk of sexual adventures in front of the hotshot women.

Hipke had a reputation of his own, without benefit of tradition, for having two of the strongest legs in smoke-jumping. He was six feet four inches tall, weighed two hundred pounds and in the off-season played ice hockey and bicycled over the mountains near his home in western Washington State. He would climb anything, from a mountain to a two hundred-foot-high Douglas fir tree—he specialized in climbing trees to retrieve parachute gear. And he worked without gloves to keep his hands tough, safety regulations to the contrary notwithstanding.

As Hipke walked along the west-flank line, he moved aside chunks of shale and scraped with his Pulaski, building up the line so it would hold rolling firebrands. He exchanged a word with Tami Bickett. Her Prineville crew was preparing to resume work, and she asked him to move ahead of or behind them. A combination of things—a gentlemanly spirit, a willingness for work and luck—kept driving Hipke farther along the fire line. He walked until he came to the last of the firefighters, about two hundred yards back from the crest of Lunch Spot Ridge. Jim Thrash, the forty-four-year-old Idaho jumper who guided hunters in the off-season, was talking with one of the hotshot rookies, Terri Hagen.

"You've got to hear this," Thrash called out to Hipke. "Go ahead," Thrash said to Hagen, "tell him what you told me."

"No . . . no," Hagen said, blushing and stammering.

Hagen had made a rookie mistake: She'd thought the different-colored hats and packs worn by the jumpers meant they were a "con crew," firefighters under private contract who are considered much inferior to smoke jumpers, especially by the jumpers.

The three were smiling over the misunderstanding when they saw Roger Roth, a mirror image of Thrash in full beard, walking toward them carrying a five-gallon cubitaner of water. In the seconds Roth took to put the water down and straighten up, the smoke column beyond Lunch Spot Ridge changed from a tracing of white to a marker line of black.

They turned their faces as one toward the smoke.

THE OTHER PAIR of Idaho jumpers, Mike Feliciano and Mike Cooper, finished eating and walked down the rocky backbone of Lunch Spot Ridge. They found a place where they could see ahead, to the bottom of the smoke column. The smoke rose from flames crackling along the lowest edge of the fire, far down toward the bottom of the western drainage. Flaming pinecones, sticks and duff tumbled toward the bottom of the drainage, the supposed safety zone.

"You know what it's going to do?" Cooper said. "It's going to back down into this draw and come up on this side at us. That's what it's going to do."

"Then why don't we get out of here?" Feliciano said. They continued to gaze at the sputtering fire.

AS DALE LONGANECKER finished eating, he heard the Grand Junction dispatcher on the radio saying that another group of jumpers, twenty-eight this time, were headed to Storm King by bus. The wind at Walker Field had strengthened, canceling parachute operations.

Longanecker walked down the far side of Lunch Spot Ridge beyond the west-flank line. Ahead, two eroded washes, which came to be called the Double Draws, ran from the top of Hell's Gate Ridge to the bottom of the western drainage. They would affect any fire moving up the drainage, but could act either as chimneys, drawing up the flames, or as a firebreak, protecting Lunch Spot Ridge. Beyond the Double Draws fire smoldered in the now-familiar spiderweb pattern. The oak had given out, and more substantial PJ and pine trees covered the slopes ahead.

Longanecker decided to flag the route for the fire line at an angle toward the bottom of the western drainage. That way, the twenty-eight fresh jumpers arriving by bus could hike into the drainage, start their own fire line and quickly link it to the west-flank line. With that doubling of effort, Longanecker felt certain they would beat the afternoon wind and hook the fire. But everything depended on being fast.

The freshening wind came as no surprise to Longanecker. At the briefing that morning at Walker Field, the smoke-jumper coordinator, Rick Blanton, had said that a cold front was on the way, though he spoke hours before Chris Cuoco would issue his verbal warning of an abrupt, severe wind. Small fires may become big ones, Blanton had cautioned.

After parachuting in, Longanecker, as smoke jumper in charge for the second load, had talked briefly with Mackey about the overall situation, and they had agreed that their "only choice" was to beat the afternoon wind to the punch by constructing the downhill fire line.

It was about 3:00 P.M. when Longanecker started across the Double Draws. The flames ahead quickened. A blanket of white smoke swept toward the top of Hell's Gate Ridge.

The Missoula jumpers finished eating and walked across Lunch Spot Ridge, following Longanecker. They found a rock outcropping and looked out, expecting to see a small patch of fire. Instead volleys of flame, dazzling even in sunshine, broke through the blanket of smoke on the slope ahead.

A backwash of embers swirled above the flames. If sparks from the backwash eddied down the slope and reached the opposite side of the western drainage, there would be fire on both sides of the gulch. That kind of fire creates its own wind. It turns small flames into a giant fireball, and the fireball races up the gulch faster than a man can run. That had been the story forty-five years earlier in Mann Gulch: A fireball had chased the smoke jumpers out of the narrow portion of the gulch and caught them on an open hillside. If the same thing began to unfold on Storm King, the "safety zone" in the bottom of the drainage would become the pathway for the fireball and turn into a death trap.

At this moment Petrilli and the other smoke jumpers had eyes only for the lively flames on the slope ahead, not for the gulch directly below them. Longanecker, beyond the Double Draws and within a few yards of the flames, radioed that he needed a sawyer and a couple of diggers to join him to get on with cutting the fire line.

The jumpers looked at each other: No one felt like rushing forward, but no one wanted to be caught holding back. Two said "I'll go," put down chain saws and stripped off saw chaps, preparing to become lowly diggers. No one took a step forward.

Longanecker called again. The flames had surrounded him on three sides and were rising higher. It was too late for sawyers, and this time he asked for the helicopter and its water bucket. The time was about 3:45 P.M.

Within minutes Dick Good and 93 Romeo appeared over the Double Draws. The water from 93 Romeo's bucket sprayed onto the flames around Longanecker in a tight pattern, indicating moderate wind at that spot on the slope, a positive sign. But there was enough fire so that the single water drop accomplished little. Good headed for a farm pond five minutes away for a refill. It bothered him to be making a ten-minute round-trip while the Colorado River was in sight seconds away, but federal regulations prohibit flying a helicopter with a water bucket over an interstate highway.

As 93 Romeo disappeared, flames blossomed in the crowns, the topmost branches, of the trees near Longanecker. Gusts of wind picked bouquets of flame from the tree crowns and tossed them to the next trees, each throw quicker and farther up the slope than the one before. The crown fire made one run for the ridgetop, then another and another. The creeping fire had cooked pine needles to incendiary dryness, speeding the fire's advance. The jumpers watching the flames remarked to each other that they hadn't seen flames move that fast since the fires in Yellowstone Park in 1988.

"Think you could outrun that?" asked Quentin Rhoades. He realized that the situation was turning into another Mann Gulch and said, without trying to cause alarm, that the book on the Mann Gulch fire would have to be rewritten after this one was over.

Petrilli took out a camera; in the viewfinder he saw a tiny human figure, Longanecker, in a closing box of flames. Petrilli snapped a hasty shot and took the camera from his eye. "That's one place I don't want to go," he said, first to himself and then out loud.

He radioed Longanecker. "Dale, I don't know if you can see it, but the fire's cooking right above you."

"Yeah, I can see it," Longanecker replied without audible emotion.

"We really don't want to come down there, and we think you should get out of there," Petrilli said.

For long seconds no one moved on Lunch Spot Ridge.

EIGHT

THE WIND BLEW past Rifle, across the Grand Hogback and on toward Storm King Mountain, pushing the cloud cover ahead of it. The sky turned blue as steel. At approximately 3:20 P.M. a wave of wind swept over Canyon Creek Estates and advanced on Storm King and its misshapen ridges, gulches and canyons. It crossed the low ridge beyond Canyon Creek Estates, where Sam Schroeder's engine crew had made the first, aborted attempt to hike to the fire two days earlier, on July 4.

The wind dipped into the western drainage on the far side of the low ridge. It crossed above the western drainage and, disturbed by the terrain, struck the slope of Hell's Gate Ridge in an oscillating wave. Light currents of air fluttered around Don Mackey and the others on the west-flank fire line, protected by oak at the midway point on the slope.

The wind broke over the crest of Hell's Gate Ridge in an erratic pattern, sweeping tossed branches into the East Canyon in one spot but too light to be felt a hundred yards away.

Leaves smoldered, trees torched. Flames crackled along the fire's ragged edges. The wind blew clouds of smoke over the top

of Hell's Gate Ridge, creating a thick haze on the lee, East
Canyon side.

At 3:23 P.M. Butch Blanco radioed the chief dispatcher, Flint
Cheney, at the BLM's Grand Junction District. Blanco noted
from his vantage point atop Hell's Gate Ridge that the fire had
begun to flare; he told Cheney he considered the situation "normal" for late afternoon. He told Cheney he needed a few supplies—ten gallons of gasoline for the chain saws and fresh AA-size
batteries for the two-way radios. Cheney noted in the log: "Lots
of increased fire behavior."

By now Dick Good in 93 Romeo had become a one-man air
show. He had hauled people and gear to the ridge, hauled gear
away and in between tried to deal with questions from Don
Mackey and Dale Longanecker about what the fire looked like
from the air. Good was not a trained aerial observer and already
had too much to do. Afterward he had no memory of radio
exchanges about the look of the fire, though others remembered
hearing the questions and his attempted answers. It was only by
luck that Good and 93 Romeo remained on the fire, well beyond
the half-day commitment made for the helicopter that morning.
The Grand Junction dispatch office had called at 3:30 P.M. to
check how things were going and, with no fire more urgent in
the district, had told Good to stay on.

Good was heading to the mountain with a refilled water
bucket, intending to make another drop near Longanecker, when
he received a change in orders. The firefighters on top of Hell's
Gate Ridge needed a water drop. Michelle Ryerson had radioed
Mackey and told him, "It's real windy, and the fire's jumped the
line." She had asked who had higher priority, she or Longanecker.

Her fire line was the last line of defense. Mackey told her to
take the water drop.

Good altered course and flew toward the ridgetop, a bit more
than a quarter mile above Longanecker's position. As the helicopter gained altitude, the west wind caught it and whisked it
over the ridgetop like a tumbleweed. Good regained full control
in the lee of the ridge and, using the wind as a brake, edged back
toward his target. He released the water with a generous allow-

ance for wind drift and watched it burst into a mist. He figured
that the drop had been wasted, but firefighters on the ridgetop,
with a better view than he below the helicopter, observed half
the water hit and knock down flames. In the next minutes that
dousing probably kept a group of firefighters from being trapped
by flames.

Good radioed Ryerson and, half apologizing, said he would
hurry back with a refill. As he flew off, he looked into the bottom
of the western drainage. Spots of fire flickered on both sides of
the deep V. Good could hear chatter on the radio about "spot
fires" and figured that everyone saw the same thing. He flew on
without reporting his sighting.

When Good returned, nearly ten minutes later, a thick column
of black smoke poured out of the bottom of the gulch. The wind
"was howling" at forty-five to fifty miles an hour, and Good
again had to fight for control of 93 Romeo. He held the helicopter far to windward of his target on the ridgetop, allowing for
maximum drift, and released the water; this time the entire load
misted.

Good took another glance into the western drainage. Spot fires
had climbed two thirds of the way up the slope opposite Hell's
Gate Ridge. Ribbons of scarlet appeared though the churning
smoke. The fire had grown beyond anything water could control.
Oh, shit, this is really bad, Good said to himself.

There was only one mission left for 93 Romeo: a rescue
pickup. Good's first responsibility was to the helicopter crew,
Tyler and Browning, working at the helispot, H2, on the top of
Hell's Gate Ridge near the jump spot. It would be insane to try
to pull people off the mountain while trailing a cable and water
bucket below the helicopter, but maybe he could manage it,
Good thought, just maybe. First, though, he had to establish radio
contact with Tyler and Browning. If nothing else, he then could
arrange to pick them up once he jettisoned the cable and bucket.

Again using the wind as a brake, Good edged 93 Romeo toward the helispot, H2. The ridgetop was clear of smoke, and he
could see many people strung out along it, but could not tell
whether they were hotshots, BLM firefighters or Tyler and
Browning.

"Rich, Rob. Hey, guys, where are you? Talk to me!" Good called on the radio, once, twice and a third time without a response. Time was running out. He aimed 93 Romeo for the meadow at Canyon Creek Estates, leaving the ridge behind.

Good barely touched down as the ground crew unsnapped the cable and bucket, freeing him for a rescue pickup. Aloft again in seconds, he spun 93 Romeo to face Storm King.

In the few moments it had taken him to jettison the bucket, Hell's Gate Ridge and its human cargo had disappeared behind a thundering column of gray-black smoke. The mushrooming cloud blotted out the top of Hell's Gate Ridge, along with Blanco's firefighters, half the Prineville Hot Shots, Tyler and Browning and the west-flank fire line with its smoke jumpers and hotshots.

The sky over the Colorado River remained a brilliant blue with a few lingering clouds, as though it were nothing but a pleasant summer afternoon. Good could fly toward the river, half-circle Hell's Gate Ridge and come to the ridgetop from the opposite direction, from the East Canyon side. It was the only chance left.

Good turned 93 Romeo away from the erupting smoke column. Once over the Colorado River, he turned into the river gorge in the direction of Glenwood Springs. He flew with the wind at his back, past the stubby outcropping of Hell's Gate Point overlooking the river, past Hell's Gate, where Blanco's crew had started from that morning, until he could look back at the East Canyon.

He flew into the smoke haze over the East Canyon and headed for Hell's Gate Ridge.

Good began calling Tyler and Browning but again heard no answer. The two helicopter crewmen must be somewhere along Hell's Gate Ridge—why didn't they respond?

Above him a curtain of black smoke ran across the sky like a squall line, pouring toward him over Hell's Gate Ridge. Its frontal edge had a curling lip.

Good felt wind pressure build on 93 Romeo as he ascended parallel to and just below the ridgetop, under the roof of smoke. He used plenty of "cross-control," leaning the craft to counter the effect of the wind. Dirt, sticks and leaves blew through his rotor wash and into the helicopter's open doorway, stinging his face. Even a momentary lull would set 93 Romeo turning cartwheels, but now he had a clear view along the top of the ridge.

Tyler and Browning were nowhere in sight. Instead Good saw a "wide, wide flame front" beginning to rise on the far side of the ridge. In seconds a rescue pickup of anyone would be impossible. "Rich, Rob, where are you, guys? Come on!" he called. The radio had become a blare of voices without names shouting orders, questions of orders and reversed orders, but nothing from Tyler or Browning.

Then, as if by magic or theatrical direction, a line of firefighters in anonymous yellow shirts and green pants appeared out of the smoke, running along the ridgetop toward the helispot, H2. They were two hundred yards from H2, then one hundred.

Behind them an enormous wave of flame arose from the western drainage and began to sweep the ridgetop, driving the firefighters before it. It swelled to a height of 50, 100 and then 150 feet. It moved faster than any human could run; everything was happening too fast. The flame wave began to break over the ridgetop, transforming the people into surfers riding the curl of a scarlet-orange wave of fire. One by one they peeled off the ridgetop ahead of the wave, heading into the East Canyon. Good saw small spot fires beginning to burn in the canyon.

The shouts on the radio died out; there were no orders left to give or reverse. Good heard a click as someone keyed a transmitter, and then a scream, a long *AAAAHHHH* without identifying name or gender. The scream ended, or more accurately, the transmission broke off, and the radio went silent.

Why the hell didn't someone scream five minutes earlier, when there'd been time to do something? Good thought. He let 93 Romeo slip away from the ridge and instantly lost touch with the wind. He flew out of the East Canyon, over the blue band of the Colorado River, and landed in the meadow at Canyon Creek Estates. One of the helicopter's ground crew, Steve Little, rushed up to 93 Romeo. Good sat slumped in his seat.

"They're gone," Good said. "They're all burned over."

∗   ∗   ∗

IN CANYON CREEK Estates Allen Bell once again went into his yard and aimed his video camera at Storm King. As the huge smoke column rose in the background, a neighbor hurriedly drove up.

"That's a nice how-do-you-do," the neighbor said, getting out of his auto.

"They've got everything available on the way out here," Bell replied.

"They're evacuating."

"Where, here?"

"They're telling everybody to get out. That's what the firemen are doing. Boy, for just one guy with a shovel on Sunday morning [July 3, the first day the fire was seen]. *Now* they're excited."

"They're not worried about this side of the development, are they?"

"I think they're going to evacuate the whole thing."

A child joined the conversation. "Not good, is it?" the neighbor said to the child as they watched the billowing smoke.

"There are firemen in there, too," the child replied.

"There sure are," the neighbor said. He drove off to set water sprinklers as a fire buffer around his house.

Part

II

NINE

WHEN THE WAVE of flame swept the top of Hell's Gate Ridge just after 4:00 P.M. on July 6, it could have marked the end for the forty-nine firefighters on Storm King Mountain. It did not; a majority survived. Natural disasters strike down one person and spare another steps away; the fate of individuals depends on anything from blind luck to a good pair of legs to history's repeating itself. Brains, the most valuable commodity in avoiding disaster, become just another item once it strikes.

For most of the four previous days, brains could have prevented what happened next, a fatal race with fire. The BLM could have followed its own directives and put out the blaze in the first days. The firefighters could have acted on their suspicions earlier on July 6 and retreated, though no place was safe after the fire blew up.

Chance took a leading role as events played themselves out, but the fire on Storm King Mountain left time for purposeful action, so character counted, too. For fourteen firefighters the next minutes would be their last and define them forever; for thirty-five who survived, the memory of what they did and what they might have done would remain with them, haunting some, resting more easily with others.

The fire became overwhelming at the most vulnerable time for the firefighters, in late afternoon, when they were dull with fatigue and at their point of maximum dispersal on the mountain. Since much depended on location, the places where people were standing, and since after 4:00 P.M. events moved in a rush, it would be well to survey where everyone was at that moment.

The forty-nine firefighters had broken down into four groups, plus one loosely joined bunch who walked between groups. The most numerous group worked at widening the fire line on top of Hell's Gate Ridge: Butch Blanco and three other BLM firefighters, Michelle Ryerson and a mixed Forest Service–BLM crew of four, and eleven Prineville Hot Shots, including both Prineville supervisors, Tom Shepard and Bryan Scholz.

Two helicopter crewmen, Rich Tyler and Rob Browning, formed a separate group at H2, the helicopter landing zone cleared that morning on the ridgetop.

There was the bunch, numbering five, who walked between the ridgetop and the west-flank line for various reasons. Sarah Doehring and Kevin Erickson, the Missoula jumpers sent by Don Mackey to check back along the west-flank line, met up with Sunny Archuleta, the Missoula jumper who had been helping out at H2, as Archuleta started down the west-flank line to rejoin Mackey's crew. They then briefly encountered two men from Blanco's BLM crew, Brad Haugh and Derek Brixey, who had been sent to widen the west-flank line near its junction with the top of Hell's Gate Ridge.

Two other groups were on or near the west-flank line. There were thirteen firefighters in one of these groups, the same number as the unlucky thirteen of Mann Gulch: Mackey, three other smoke jumpers and nine Prineville Hot Shots. They were at the farthest extension of the line, nearly seven hundred yards from the junction with the top of Hell's Gate Ridge, starting back to work after a lunch break.

The last group—nine smoke jumpers, including Dale Longanecker and Tony Petrilli—had eaten lunch on Lunch Spot Ridge, the major spur ridge running off Hell's Gate Ridge, and after eating had walked ahead in the direction of the Double Draws, the eroded gullies beyond the crest of Lunch Spot Ridge.

That was the situation at 4:00 P.M., moments before everything changed.

The nine jumpers who had walked across Lunch Spot Ridge had been transfixed by flames dancing in the crowns of Douglas fir and ponderosa pine trees beyond the Double Draws. All forty-nine firefighters on Storm King Mountain had seen smoke rising from this place during the afternoon, their view partially blocked by Lunch Spot Ridge. Why the smoke column caused no greater concern is a question many asked themselves afterward. A look-out on Hell's Gate Ridge might have been able to give an earlier warning, but not by much. No one was in doubt that fire caused the smoke. At the crucial moment, when the smoke signal became an inferno, there were witnesses aplenty.

Longanecker, the line scout, had walked ahead of Petrilli and the other smoke jumpers, who, from an outcropping on Lunch Spot Ridge, watched as flames surrounded Longanecker on three sides. Longanecker was standing at midslope on the far side of the Double Draws.

The flames flapped and soared around him. Longanecker felt no immediate threat—he planned to retreat toward Lunch Spot Ridge before the fire completely encircled him. But the race to hook the fire, he realized, had been lost to the wind. He had come to that conclusion the moment Dick Good and 93 Romeo swept away from his location to handle the flare-up near Michelle Ryerson's group on the ridgetop. "We have a real bad situation here," Longanecker heard someone say over the radio. Indeed they did.

But even so, Longanecker stood rooted where he was, hypnotized by the flames, and the other jumpers on Lunch Spot Ridge stood spellbound where *they* were. The spectacle unfolded before their eyes. Flames danced into a sky shimmering with sunlight. Clouds of white smoke rose into an otherwise nearly cloudless sky. A blowy, beautiful summer's day swelled to fullness as the earth joined the heavens, until a movement in the deepest part of the western drainage broke the spell.

A jet of flame shot upward and then another, seeming to spring from nowhere. Piles of dead brush, branches and tree trunks ignited. Living brush, tinder-dry from drought, took fire. Darts of flame transformed into bonfires, which merged into a single, expanding flame front. A booming wind raced up the western drainage and struck the flames, pressing the telltale smoke column nearly flat to the ground.

Muscular strands of scarlet flame appeared through the smoke. The fire drew back to renew itself, taking in oxygen, and the smoke covered the flames; then the fire surged forward, and again ribbons of flame came into view.

As the smoke jumpers swung their eyes in near-unison to the sight, recognition struck home. They had fire below them on a mountain, and no ordinary fire: It was a blowup. There is no agreed scientific definition for a blowup, but it is unmistakable in appearance and overpowering in effect. The Forest Service has a working description of the event: the rapid transition of a fire burning in debris and litter to one involving all available fuel, from the ground to the tops of trees. But this falls short of describing the majesty of the occasion.

A blowup is one of nature's most powerful forces, equivalent to a mighty storm, avalanche or volcanic eruption. It can sweep away in moments everything before it, the works of nature and of humankind, and sometimes humankind itself. It is destructive, but neither good nor evil; it goes where wind and terrain take it.

Blowups happen every fire season across the West when wind, fuel, dryness and terrain come together in the right combination and meet with a spark. The blowup stokes itself by creating its own wind, the heat drawing cooler air by convection. If it happens in a gulch, as is common, the sides of the gulch—in this case the western drainage—act as a chimney and compress its energy. The flaming tempest can send a smoke column to a height of forty thousand feet or more. The blowup may die out once the gulch is burned or move on and reduce thousands of acres to ash. The blowing-up, in any case, is over in minutes.

A blowup is as different from the smoldering, four-day-old fire on Storm King as a hurricane is from a summer squall. But it was the sputtering fire on the upper slopes of Hell's Gate Ridge that caused the blowup in the western drainage. Flames, slowly backing down the slope of the ridge, marked a thin, broken trail into the bottom of the drainage. This is what a lookout would have seen, and what the jumpers who walked across Lunch Spot Ridge did see; it was not a terribly threatening sight. The flames advanced one tree or clump of brush at a time, slowly but surely. Flames also made downhill leaps as wind eddies scattered sparks toward the bottom of the V.

The west wind created the eddies when it struck the ridges of Storm King and partially turned back on itself, just as flowing water turns in eddies when it passes rocks or other obstacles. The eddies carried aloft fistfuls of burning duff, that is, decayed leaves, twigs and other matter. Most likely a combination of flames backing down the slope and embers spinning in eddies ignited the initial, small fires in the bottom of the western drainage.

The west wind fanned those fires into a blowup, but to do so it first had to curl around a mountain and become a south wind. The west wind entered and was intensified by the gorge of the Colorado River, a natural wind funnel, in a phenomenon known as a venturi effect, named for the nineteenth-century physicist G. B. Venturi, who discovered that a throatlike, constricted tube will increase the velocity of fluids. Once in the river gorge the surging west wind quickly found an escape valve and turned at a right angle into the mouth of the western drainage, transforming itself into a south wind. It now raced up the narrow V of the drainage, which further compressed and accelerated it.

As the south wind sheared around a turn a half mile up the western drainage, it came upon the spot fires. When the wind struck the flames, they exploded. A thunderous roar rose from the gulch. The transition from a "normal" fire to a blowup took seconds.

Minutes before, the forty-nine firefighters had been ordering gasoline for their saws, taking the fire's photograph and making a game of the similarities to a legendary killer, the Mann Gulch fire. Then disaster took its unmistakable shape, and the firefighters, almost as one, began a race for their lives.

WHEN TONY PETRILLI saw the fire blow up in the drainage below him, he radioed Mackey on the west-flank line.

"The fire's spotted across the canyon, and it's roaring," Petrilli said, struggling to keep his voice calm.

"Is it across *the* main canyon?" Mackey asked.

"Yes, it's across *the* main canyon. It's rolling, and we're getting out of here," Petrilli said. He told Mackey he was coming back over Lunch Spot Ridge, toward the west-flank line. Whatever else happened, it was certain that the blowup would gut the western drainage, the designated safety zone.

Petrilli took a last look, and a before-and-after picture became etched in his memory. The spot fires, the size of campfires when he had called Mackey, had expanded into a flame front that had already raced fifty yards up the drainage.

Petrilli counted hard hats around him—six, including his. The line scout, Longanecker, was going to have to find his own way out. The two Mikes, Cooper and Feliciano, who had been nearby, had begun their own retreat and now joined Petrilli and the other smoke jumpers. As the eight jumpers headed back over the crest of Lunch Spot Ridge, they were struck by a wild, gusting wind, the kind that precedes a violent storm. Several tightened chin straps to keep their hard hats from blowing off.

Wisps of smoke fingered through the pine and juniper trees as Mackey, eyes red-rimmed and tearing, came up to the crest of Lunch Spot Ridge. Seeing Petrilli and the others coming up from the opposite side, Mackey waved and shouted, "Go up, there's good black farther up." Petrilli stopped next to his friend, and he, too, shouted, "Up the ridge." Mackey scanned faces as the jumpers went past him. "Where's Longanecker?" Mackey asked. Petrilli had last seen Longanecker gazing at flames beyond Lunch Spot Ridge. Mackey grabbed his radio: "Dale, are you okay?"

"Yes," replied Longanecker, never a man of many words. Longanecker watched a fireball run up the slope of the western drainage opposite Hell's Gate Ridge, away from him. It was time for him to retreat back across the Double Draws toward Lunch Spot Ridge.

What of the others back on the west-flank line? Longanecker and Mackey asked each other. That group, nine hotshots and a handful of jumpers, were in a bad spot, the two supervisors agreed. During the afternoon Mackey had several times referred

to the "people in the brush patch," meaning whoever was working in the tunnel of oak, and the phrase now took on an ominous
meaning; being "in the brush patch" meant being blind to a
blowup.

Mackey told Longanecker that he was going back to "check
on them" or "to get those people out," Longanecker said later,
unsure exactly which phrase Mackey had used, but in no doubt
about his intentions.

As Mackey and Longanecker talked, Petrilli and the other
jumpers continued up Lunch Spot Ridge, at least one of them
nursing doubts. Quentin Rhoades wondered how good Mackey's
"good black" could be.

Rhoades knew that they would find a blackened, cleared zone
at the top of Lunch Spot Ridge, because that was where they
had camped the previous night. But so what? A lot of burnable
stuff lay between them and the old campsite.

And what was so good about that black anyway? Minutes earlier Rhoades had witnessed flames ripping through what should
have been "good black" beyond the Double Draws. The fire
could do the same thing on Lunch Spot Ridge, trapping him and
the others.

Rhoades had made up his mind to retreat along the west-flank
line, a familiar path and one that led away from the erupting
smoke column. He halted when he reached the line.

The spectacle before him was staggering: In eight years of
fighting fires Rhoades had never seen anything so powerful. The
blowup had moved up the western drainage and now was below
the west-flank line, running on a course parallel to it. The west
wind caught the smoke column as it rose out of the drainage and
had begun to draw it like a canopy over the mountainside.

A retreat along the west-flank line would be suicidal, but
Rhoades had to move. In the moment of decision Rhoades, the
law student, trusted Mackey, the student of the woods, and
Rhoades headed for the "good black" at the top of Hell's Gate
Ridge.

The slope steepened, and Rhoades became bent over, his free
hand—the one not clutching a chain saw—brushing the ground.
His leg muscles burned, and his mouth turned cotton-dry. He

caught up with the other jumpers and slowed behind one, Sonny
Soto, who was laboring. Rhoades felt a twinge of pity, but then
in his mind's eye he saw his wife, Suzanne, and five-month-
old baby, Rachel. He felt a knot tighten in his stomach. Feeling
no more sympathy, he accelerated past Soto and several other
jumpers.

After three hundred yards Rhoades realized that he still carried
the twenty-five-pound chain saw. He set it down, taking an extra
moment to find a spot where the saw might escape flames. Leaving it behind was "like running up a white flag." He had abandoned everything—his plan for retreat, his ties to his fellow
jumpers and lastly the tool that gave him a purpose on the mountain.

He recognized the knot in his stomach for what it was: the
fear of death. But his mind refused to picture real death, the final
moment of suffocating flame and smoke. Instead he saw his death
as from a distance, an event causing trouble for others dear to
him, his wife and daughter, but somehow not touching him. "I
realized our plight was all too real," he said later. "I just moved
out as fast as my stubby little legs would carry me."

Petrilli let the jumpers pass and then started up Lunch Spot
Ridge behind them, figuring that Mackey was on his heels. He
stopped, as Rhoades had, at the west-flank line, but gave the
scene only a brief glance. This is one place I don't want to go,
he thought once again, as he had when confronted by flames
moments before on the other side of Lunch Spot Ridge.

Windswept branches pointed up the ridge like signposts, and
he followed them.

As Petrilli struggled upward, an old familiar song began to play
in his mind, one that had chased him up other mountains with
other fires behind him, the driving rhythm and lyric of "The
Battle of New Orleans": "*We ran through the briars, we ran through
the brambles, we ran through the bushes where a rabbit couldn't go.*"
This time a less heroic battle scene came to mind: He and the
other jumpers were nothing like American frontiersmen outwit-
ting British troops; they were like mice scurrying from a cat.

After fifty yards Petrilli glanced back for Mackey. The landmark of the west-flank line was fast disappearing under a cloud

of smoke. There was no sign of his friend. Oh, shit, Petrilli said to himself with resignation, and he paused for a moment. I'd better go, he thought.

Ahead of him one jumper carrying a gasoline container halted and said out loud, as though giving himself an order, "I'm setting the gas down right here." Petrilli took the cue and set down his chain saw. The move made "the pucker factor jump a notch" but also made him worry, as Rhoades had moments before, for the safety of the saw.

Blood hammered in Petrilli's head; shouts blared over his two-way radio; the fire thundered. All sounds became equally meaningful and meaningless, until in a burst of clarity a single voice broke through on the radio: "There's a spot below you guys. Get the hell up here *now*!"

It was Kevin Erickson, Petrilli's fellow Missoula jumper; he sounded faraway, calling someone else. But the message could be for anyone on the mountain.

The "good black" promised by Mackey lay just ahead.

Keith Woods and Sonny Soto, their legs cramping from effort and dehydration, lagged behind the others. "Come on, guys, don't stop!" several jumpers called as they passed them by. Woods's thigh muscles had seized up, then the muscles of his upper calves. He tried to pull himself forward, but his arms started to cramp. He stopped to let the spasm pass.

The jumpers came at last to a broad, burned-over area a hundred yards below the top of Hell's Gate Ridge. They had climbed five hundred yards from the point where they had first seen the blowup. The "good black" was thinly scattered PJ, well scorched, the best safety zone they had seen all day. A thick, unburned patch of oak grew nearby, along one side of Lunch Spot Ridge, a possible source of trouble. Smoke and ash had followed them, but no flames as yet.

Everyone halted, chattering nervously, with Woods and Soto sixty-five yards below the others. There was "an overwhelming feeling of relief," Rhoades noted, "like we were going to make it."

They cleared patches of earth to set their fire shelters, floorless aluminum pup tents that reflect heat; firefighters call them "Shake 'N Bake" or "turkey cookers." The wind set the metal rattling.

Rhoades thought that they looked like a party of gravediggers.

Petrilli glanced at Rhoades, the first time he had looked another jumper in the face since the blowup. Rhoades was so covered with soot that even his teeth were black. I must look the same, Petrilli thought.

As Petrilli prepared to enter his fire shelter, he checked his watch. It was 4:24 P.M., twenty minutes since the blowup. He radioed Mackey.

"We're in the black on the ridge, and we're going to shelter up," Petrilli said. Nobody answered. Petrilli drew his shelter over him, put his face into the freshly disturbed earth and waited for whatever came next.

TEN

Don **MACKEY FOUND** himself alone on Lunch Spot Ridge. He had finished his exchange with Dale Longanecker with the promise to check on the "people in the brush patch," the hotshots and smoke jumpers on the west-flank line. They must be able to see the billowing smoke for themselves; they already could be in retreat. A radio warning to them, surely, would satisfy duty.

"Okay, everybody out of the canyon."

Mackey made the call a few minutes after 4:00 P.M., a time when he very likely was standing on Lunch Spot Ridge. He had moments to decide what route to take and no time for second guesses. The safe way was to follow Petrilli and the others up Lunch Spot Ridge. The west-flank line already had fire below it.

Mackey took a first step, his boots grinding into loose shale on the side of Lunch Spot Ridge; he wore new White Boots, $300 a pair, the smoke-jumper model with high heels that dig into a mountainside. He descended on or parallel to the west-flank line in long, strutting steps, almost leaning backward.

After the first thirty yards the slope evened out, and he could accelerate.

Of the many people interviewed for this book who knew Mackey well, not one expressed doubt or surprise that he willingly risked his life for others. That was Don, he always did his duty and more than his duty; he held people together.

The Prineville Hot Shots were mostly youngsters in their twenties and would stick together like family, Mackey knew from his days as a hotshot. Four of the hotshots on the west-flank line were women—Tami Bickett, Bonnie Holtby, Kathi Beck and Terri Hagen. Mackey had a special link to women, starting with his sisters.

But what about the smoke jumpers? Two of his best friends, Kevin Erickson and Sarah Doehring, were somewhere back along the west-flank line as well.

Later Mackey's parents, Bob and Nadine, would struggle to understand his decision. "He was a flirt, he really liked women, and he did it for the women," said Nadine, lashing out against a loss she could not accept.

"No, Nadine," Bob replied calmly, putting rhythm into the words, "he did it for all of them, the men *and* the women."

Mackey had been an endurance runner in high school, middle distances and not sprints, but his best athletic performances came in the mountains, hiking dawn to dark. He could "power-hike," something between a walk and a jog, all day on broken, uneven terrain. He saved himself by slowing on rises and made up time by accelerating on downslopes.

On the west-flank line it took concentration simply to stay on his feet. After the fire Jack Cohen, a jogger and member of a team of scientists studying the behavior of the South Canyon fire, made a series of attempts to run, walk and jog along the line. When he tried to hurry his pace, he wound up tripping and falling.

"You're focused on your next step; you almost feel the world closing in," Cohen said afterward.

The next spring after the fire Bob Mackey went over the ground as well, retracing Don's path step by step, remarking where he thought Don would have jogged, slowed or power-hiked. Don and Bob shared the same physique—long, strong legs; Bob covered the ground swiftly and with apparent ease.

While Don Mackey raced to catch the people in the brush patch, the blowup chased him. It churned in underbrush and sent flames into tree crowns. Smoke and embers began to skyrocket out of the western drainage.

The Prineville Hot Shots, it turned out, had seen the blowup for themselves and begun a retreat before Mackey reached them. Jon Kelso, the hotshot who was an ex-Missoula smoke jumper, had radioed Tom Shepard, the Prineville superintendent on the top of Hell's Gate Ridge, at 4:04 P.M. Kelso said, in a calm, steady voice, that he could see flames in the bottom of the western drainage.

"Tom, there's a spot across the draw," Kelso said.

"Get out of there," Shepard snapped. "Abandon the fire line. Come up here."

"We're on our way," Kelso replied.

Kelso and the hotshots with him—three women, Bickett, Holtby and Beck; and four men, Scott Blecha, Levi Brinkley, Doug Dunbar and Rob Johnson—apparently gave no thought to Lunch Spot Ridge as an escape route. They turned back the way they had come.

The four others on the west-flank line—the smoke jumpers, Eric Hipke, Jim Thrash and Roger Roth; and the Prineville rookie, Terri Hagen, who had been joking with them—saw the blowup at virtually the same time.

The column of black smoke was the excuse they had been looking for all day to get out of the oak brush. Being farthest to the rear, they became the leaders of the retreat on the west-flank line when they turned back.

The first low rise slowed them enough for Kelso and the others with him to catch up. Now numbering twelve, the firefighters proceeded on as a group, squeezing together on rises and spreading apart on downslopes like an accordion. Mackey was well behind them at this point.

At the crest of the low rise they glanced over their shoulders and caught glimpses through the brush of the erupting smoke column. Hipke noted a "big old wall of flame" following them.

"Jesus, look at that!" one firefighter said.

"Any place to shelter up?" several others called out.

"No, no," others answered. Each time they took a backward glance, they quickened their pace.

The blowup spread onto the sides of the drainage, heading into a bowl-shaped area a half acre in size on the side closest to the firefighters. The bowl was crammed with brush, providing the blowup with an early energy boost. Nearly all the fuel in the bowl was consumed; the up-gulch sides of trees were scorched, but not the down-gulch sides, indicating that the wind whipped the fire past the trees, allowing them to burn only their lee sides. The blowup swept through the bowl and onto a broad bench, sending embers ever higher up Hell's Gate Ridge.

The blowup moved at its slowest rate in the first minutes, about two miles an hour by later calculations, allowing the firefighters to stay ahead of it. The twelve smoke jumpers and hotshots hiked along at about three miles an hour, an average walking speed on flat ground but a fast clip on the rugged fire line; Mackey, behind the others at this point, moved quite a bit faster than three miles an hour.

The blowup next confronted an obstacle, a sheer rock wall of pinkish shale that rose to a height of forty feet. The rock face extended from the side of the drainage opposite the firefighters. It forced the blowup into a right-angle turn, in the direction of the firefighters.

The blowup then split, one portion completing a dogleg turn around the rock face to follow the drainage, while another portion continued straight on for the firefighters. The dogleg turn set flames spinning, carrying sparks farther up slopes.

Even before the blowup reached the rock face, it had begun to "hook and run," sending offshoots up Hell's Gate Ridge. Those offshoots became the wave of flame that spread along Hell's Gate Ridge.

The twelve firefighters now began to ascend the spur ridge, the last barrier before the final approach to the top of Hell's Gate Ridge. Until now Thrash and Roth, who were in the lead, had set an accelerating pace. Thrash, at forty-four the oldest jumper on the mountain, was a noted hiker; he had gained local fame, or notoriety, for his method of testing rookie jumpers. "I like to walk 'em till they puke," he often said.

But the spur ridge slowed him and the others to a pace of barely over a mile an hour. The blowup, meanwhile, more than doubled its speed as it turned uphill following them; it accelerated to about five miles an hour, five times faster than the people. The distance between them and the blowup began to close at an alarming rate.

Hipke saw flames "curling and whipping" behind them as the fire hooked up the slopes. Hipke was habitually tardy for appointments and now had the familiar feeling of having to use his exceptional speed to make up for lost time. Third in line, he began to look for a place to pass Thrash and Roth.

Hipke shoved the backpack ahead of him and shouted "Go! Go!"; others behind him did the same thing. As the west-flank line passed through a clump of trees, the cup trench in its center split into parallel paths and then rejoined after a few steps. It was no more than a blip, but it was Hipke's chance. Thrash and Roth took the same fork, leaving the other one wide open.

I could do it, Hipke thought, but . . .

He checked himself. Busting past fellow jumpers would be a brazen move. And there was enough space to pass one jumper, not two. The opportunity vanished as Thrash and Roth came back onto the undivided path; the firefighters continued in single file.

The final yards up the spur ridge offered an improved view of the blowup; Hipke turned and saw an unexpected sight. The group had bunched together, and there, in the last position, was Mackey. What's he doing with us? Hipke wondered.

Mackey's round face had twisted into a grimace, lips parted and teeth bared. Hipke recognized the look of extreme physical exertion. Their eyes locked, and Mackey's pleaded, Go! Go! Go!

Mackey had covered the ground, about a quarter mile, roughly twice as fast as the others, maybe a touch faster. That meant an average speed for him of six to seven miles an hour, or a pace of a mile in eight and a half to ten minutes, a slow jogging speed on flat ground but a test of even a fine athlete on the west-flank line.

He could sense the fire, though, without taking his eyes off the path. The sky darkened as the blowup shut out the sun. The fire rumbled with a constant thunder that joined the pounding of blood in Mackey's head, the rasp of his breath and the hammering of his boots. The south wind driving the blowup met the prevailing west wind at midslope, where Mackey was, mixing in a "shear layer" of swirling sparks, smoke and debris. He began to breathe in large quantities of smoke.

Mackey, expecting to find the other firefighters where he had left them, near Lunch Spot Ridge, must have wondered if he was on a fool's errand when he found that stretch of line vacant. He crested the first low rise, jogged down its opposite side, then moved swiftly along the relatively easy stretch leading to the spur ridge. He pounded past the spot, now vacant, where Roth had wrestled with the smoldering log.

Mackey hit the spur ridge in a power hike. Farther ahead, the twelve firefighters on steeper ground slowed to a crawl. Within moments Mackey glimpsed backpacks and blue hard hats in the line ahead of him: He had done it, he had caught up. But that was only half the job.

If *he* could catch them, so could the blowup. He was too winded to yell and too far from the leaders to be heard if he did. He looked up, saw Hipke, and his eyes said everything: Go! Go!

The thirteen firefighters crested the spur ridge, each with a twisted face and whistling breath. The top of Hell's Gate Ridge loomed ahead, seemingly close enough to touch, promising safety. It was a mirage. Dips and rises added many yards to the apparent distance.

The firefighters continued to carry their packs and tools, including chain saws. If they had to set up fire shelters, there was nothing like a Pulaski for clearing a spot.

Mackey caught his breath and reported by radio, "We're coming up to a good spot." Shepard, on the ridgetop, recognized Mackey's voice and remarked its urgency.

At about this moment two Missoula jumpers, Sarah Doehring and Sunny Archuleta, were standing just off the top of Hell's Gate Ridge, overlooking the west-flank line. They began taking photographs.

Doehring and Kevin Erickson had walked back along the west-flank line, following Mackey's direction, to check for rolling embers. When they had reached Roth, the smoldering log and the cubitaners of water, Erickson had offered to help carry the water. But Roth had said not to bother, he could handle it.

Doehring and Erickson then crossed the spur ridge and ran into Archuleta on the flattish stretch of line before the hundred-yard climb to the ridgetop. Archuleta was on his way to rejoin Mackey and the others after helping the helicopter crew at H2. He thought that the pointed stobs along the fire line looked dangerous, like "punji sticks," but otherwise was unalarmed. As the three of them said hellos, an excited swell of voices came over the radio.

"It's down the canyon now! . . . Hey! Hey!"

Doehring looked over her shoulder. "They're right, things are cooking up," she said.

Then Archuleta heard Mackey's familiar voice over the radio: "Okay, everyone out of the canyon." That made up Archuleta's mind: He turned, with Doehring and Erickson on his heel, and began to climb to the top of Hell's Gate Ridge. Halfway up they met two other firefighters, Brad Haugh and his swamper, Derek Brixey, standing next to a pine tree, one of the few on the slope, which later became a landmark known simply as The Tree.

Archuleta asked Haugh and Brixey if they had heard the order to pull back. What order? they asked, taking out earplugs and shutting down their chain saw. Archuleta reported Mackey's order to leave and then, as if to illustrate the point, marched on. Archuleta tried to tell himself that he was hurrying to find a good place to observe the fire, but he knew better. He wanted out of a death trap.

Erickson remained at The Tree with Haugh, waiting for Mackey and the others to show up.

When Archuleta, with Doehring and Brixey following, reached the top of Hell's Gate Ridge, the scene was chaotic. The Prineville Hot Shots and the others on the ridgetop darted in every direction—"like a pack of wild dogs," thought Brixey, who disappeared into the confusion.

Archuleta and Doehring walked a few yards and then picked

their way down the slope until they found an opening in the oak and PJ. Blue hard hats appeared on the west-flank line about two hundred yards away. No faces were visible at that distance, but Prineville Hot Shots and a few jumpers wore blue hats; it must be them—most jumpers wore hats of other colors.

Here they come, that's great! Doehring said to herself, but wondered where "our guys" were, the rest of the smoke jumpers who had been on the west-flank line.

As Archuleta and Doehring watched, a jagged crescent of flame arose behind the spur ridge and framed the bobbing blue hats. Tongues of flame probed the air. Smoke cast a dark shadow over the human figures. The spur ridge acted for the moment as a barrier against this portion of the fire, but it also blocked the firefighters' view of the flames rising behind them. The blue hats progressed at a regular pace.

Archuleta gauged the distance between the flames and the hard hats and their relative speeds, and made a mental calculation: Those guys are fucked, he thought. He asked Doehring if she had a camera. She dug one out of her pack.

Archuleta knew from experience how vital photographic evidence could be at a moment like this one. Three years earlier, in May 1991, he had been assigned the routine job of videotaping a training parachute jump near Missoula and had wound up recording the most recent accidental smoke-jumper fatality. His film had been crucial to the Forest Service investigators' conclusion that the accident had resulted from human error, the failure of the smoke jumper, Billy Martin, to deploy his primary and reserve parachutes in time.

Archuleta took the camera from Doehring and snapped two quick shots.

The first photograph appears to show three people bunched together on the west-flank line, perhaps actually touching, with the crescent of fire behind them. The figures can be made out in extreme enlargements, but no faces are visible. Though others must be in the photo's frame of reference, they are hidden by brush and terrain. Staring long enough at the enlargements brings out other shapes, suggestive of human limbs. But they could as easily be oak branches.

The three discernible figures are coming along the fire line after it crosses the spur ridge. They have a substantial distance to go—580 feet, to be exact—to the top of Hell's Gate Ridge.

The photo raises a question: Why are the three figures so close together? The fire line was free of stobs only in the eighteen-inch-wide cup trench at its center, too narrow for three people to hike abreast—the group in fact marched in single file. The three people, then, came together for a temporary purpose. What was it?

The firefighters on the west-flank line knew they were being chased by a blowup. These three could be grabbing each other's elbows and offering a word of mutual encouragement after the difficult hike over the spur ridge.

Or one of the three might be injured and receiving attention, sandwiched between the other two. There is no hard evidence for this explanation, however, beyond the unusual grouping.

Two of the figures appear to be wearing Prineville blue hard hats. A long empty stretch of fire line is visible behind them, so they evidently are the last in line. Investigators at first thought Mackey was in the photo, between the two, but a more precise enlargement made several years later raised doubts. There is a third shape, a yellow object partially hidden behind one of the figures. It could be someone's fire shirt or something else; there is no way to be certain. Mackey may have moved farther ahead in the line.

Mackey, a supervisor who had just rejoined the group, would have talked with those around him. His radio message, the one about "coming up to a good spot," offers a clue to what might have been happening at this point.

Shepard was absolutely certain of the wording, so what did Mackey mean by a "good spot"? Good for what? Those words imply something more precise than the presumed safety of the ridgetop.

Mackey might well have been talking with the others about deploying fire shelters. The group had been talking about shelters before Mackey caught up to them. The question should have come up again about this time—the blowup was closer, the smoke jumper in charge had shown up, and they all were catching their breath after a steep climb. The people were photographed in a place that would have made a poor deployment site, a narrow path through dense brush in a saddle where fire would be intense. Mackey might have been reporting a decision to move farther along to a "good spot" to deploy.

By the time Archuleta took his second photograph, the people had moved forward, disappearing in the oak. Eventually several fire shelters *were* deployed. So this theory covers the facts and makes sense, but whether it is true cannot be known.

Archuleta handed the camera back to Doehring, who snapped two shots of her own, though she later did not remember having done so. Both her photos show Archuleta staring trancelike at flames as bright as his orange hard hat. No people are visible on the fire line for certain. Doehring did remember her next action: She stuffed the camera down her shirt, said to herself, Holy shit! I'm out of here, and left.

When Doehring regained the top of Hell's Gate Ridge, she saw firefighters still carrying chain saws and gasoline containers and shouted, "Drop that stuff, we've got to get out of here!"

Archuleta, meanwhile, watched hypnotized as the fire rolled toward him. He started to say something to Doehring and realized she was gone. He finally broke away.

When he regained the ridgetop, he headed toward the west-flank line, telling himself, At least I can keep anyone else from going down there. Archuleta set his Pulaski in the ground like a staff. As firefighters hurtled past him, pursued by a wave of flame, Archuleta waved them on.

The roar of the fire became deafening. Archuleta hollered at a passing figure, "Are you the last?" and heard a muffled "Yes!" The air around him turned orange-yellow, as though the sky itself had caught fire. Sparks fell at his feet. Radiant heat pulsed out of the depths of the western drainage, forcing him back.

You stayed too long, Sunny, he told himself. You've made your last mistake.

THE TWO FIREFIGHTERS who had lingered at The Tree, Erickson and Haugh, figured they had time to spare. The slope above The Tree, which Archuleta, Doehring and Brixey had just climbed, rose at a daunting fifty-five degrees—so steep it was difficult to see the crest of Hell's Gate Ridge. But a scramble of a mere two hundred feet would take them to the top. The fire, for the moment, remained far off.

The two chatted. Erickson asked, "Does this oak burn hot when it goes?"

"It burns like a bastard when it's dry," Haugh replied, "say, after a frost in the fall."

"Shouldn't we be getting out of here?" Erickson asked.

"Let's hold on," Haugh said, remarking that more people from the west-flank line should be showing up any minute. "Let's give them a little support, maybe grab a saw or lend a hand."

The first hard hats came into view, bobbing blue dots punctuated by one black dot. Erickson recognized his brother-in-law's black hat: Mackey had moved up one position, to second from the rear.

If they hustle now, Erickson thought, everything will be fine. They must be puffing from the climb over the spur ridge; they would catch their breath in a moment and step up the pace. When they reached The Tree—they were no more than 150 yards away—everyone could rush for the top of Hell's Gate Ridge together.

The air began to vibrate, as though a jet airplane were on approach. Erickson grabbed his radio. "Come on, guys, get moving. You've got fire down there!"

The first of the firefighters came within earshot, and Haugh began yelling to hurry. Unaccountably, at a distance of about a hundred yards from The Tree, the line of hard hats halted and then snaked forward again after a pause.

As Erickson watched, a tiny flame sprang to life well up the western drainage beyond the barrier of the spur ridge. The flame swelled to the size of a truck; nothing stood in the way of its sweeping up the side of Hell's Gate Ridge.

Erickson grabbed his radio a second time: "There's a spot below you guys. Get the hell up here *now*!"

The firefighters moved in orderly fashion with chain saws on their shoulders and Pulaskis in their hands. The leaders approached the steep hundred-yard rise.

They had time, Erickson figured, they could make it. And

what a great picture! Don and his crew storming out of the gulch with fire boiling up behind them. Erickson had a camera wrapped in a sock in his pack and asked Haugh to pull it out for him. Don was going to love this, Erickson thought. Something to show the folks at home.

Erickson slipped the lanyard over his head. A hand shook his shoulder. "Let's get the hell out of here!" Haugh yelled.

Haugh would later swear that at this moment he saw a bearded figure, Thrash or Roth, close enough to touch, saying words to the effect of "Deploy, deploy." Subsequent inquiry raised strong doubts that he was ever that close to Thrash or Roth. He maintained the claim, however, though it cost him some credibility in the fire community.

Haugh turned and took off for the ridgetop.

Erickson put his eye to the camera: "It was bright orange, just bright orange." Erickson snapped the shutter, hands trembling, and fumbled the camera down his shirt front, missing several times. Later it was discovered that he had used up the film or exposed the same frame twice—something had gone wrong—and there is no photographic record of what he saw, or what Haugh might have seen.

Erickson glanced down the slope and saw "a guy with a beard" coming toward him, too far away to make out his face. Erickson turned and began to climb, telling himself, Don't run, don't run, they'll think you're a sissy. The fire bellowed behind him. They'll think you're a goddamn pussy! Erickson's legs pumped as never before.

He was a few yards from the top when he felt a slap across his arms and shoulders, as though a blowtorch had passed over him. His body arched, elbows drawn back, as heat penetrated his fire shirt and scorched his arms. He heard screams, his own for sure but others', too.

Then he was on the top of Hell's Gate Ridge, finally able to run. He crashed into brush on the opposite, East Canyon side and clattered down through loose shale, dodging trees. Haugh was ahead of him, sliding on his rear end, feet forward, toward what looked like a shear drop-off.

The slope was so steep that Haugh came to his feet by flexing his knees. Haugh lurched forward and slammed into a tree a few feet short of the drop-off; he was stunned but erect. Erickson came to a halt beside him, and they looked back.

An orange-yellow fog, its insides writhing with flames and gases, had followed them over the crest of the ridge. As they watched, a giant human form materialized out of the fog and bounded toward them. They could not make out who it was, except that it had to be someone from the west-flank line. The figure had legs as long as stilts and fingers of unbelievable length, even for a giant. Its bare head trailed smoke. It stumbled to a halt in front of them.

"Where is everybody else?" the giant said. "What happened to everybody?"

They had no answer.

THE FIREFIGHTERS ON the west-flank line came to a full halt near the foot of the final hundred-yard climb to the top of Hell's Gate Ridge at approximately the same time as the Doehring-Archuleta photographs were taken. At this point the line widened enough to accommodate just a couple of fire shelters, not thirteen.

The blowup had become a "wide coherent fire burning the entire fuel complex," fire scientists later said. It swept through shiny green oak, dead branches, any fuel that stood in its path; no place was safe, not the black and not the green.

Thrash stepped to one side of the cup trench and uttered a single word—"Shelter?"—halfway between a question and a statement. Roth stepped to the other side, and a path opened between them. This time Hipke did not hesitate.

I can't believe these guys want to ride it out in shelters, Hipke told himself. He had only a nodding acquaintance with Thrash and Roth and knew none of the hotshots; so nothing held him back. He was going to run for it.

Jumpers are taught to count on themselves alone when things go bad, but the question "How bad?" is never an easy one. Firefighters have their own guideline: When the worst becomes unmistakable, they call it FEAR, or Fuck Everything and Run.

Hipke spurted between Thrash and Roth and began, literally,

to skate up the ridge, elbows flying and legs digging in as though on ice skates. Ahead at The Tree he saw two figures, Erickson and Haugh, waving and shouting, "Drop your tools! Run, run!"

Hipke checked over his shoulder. Roth was coming along, leading several other people, but he had fallen a dozen or more yards behind Hipke. The halt had been costly. Hipke had expected that at least a few people would follow him. He locked eyes with Roth and saw a look of deep concern.

The western drainage was filling with fire; Hipke resumed his scramble, using a free hand to pull himself along. The slope steepened to fifty degrees. Hipke hardly noticed The Tree as he clambered past it, except to mark that Erickson and Haugh were no longer there. The slope steepened to fifty-five degrees. An inner voice screamed at Hipke, Get out! Get out! Get out!

The air turned dark red. Radiant heat pulsed from behind and from his right side. A tidal wave of sound and whirling embers engulfed him. He dropped his Pulaski—it later was found twenty-seven yards below the ridgetop—and raised gloveless hands to cover his ears; he was wearing his hard hat backward, and the bill protected his neck. Believing that his fire shelter would make a better shield, he took it out, hooked a finger in the pull ring of its plastic case and yanked. The shelter spilled out of the case and slipped from his hands. He staggered on.

I can't believe this is happening, he said to himself. I'm getting burned! This doesn't happen; you don't get caught, you just don't.

He sensed something huge swelling up behind him. How did the fire move so fast? Was it there now?

The rim of the East Canyon that adjoins Hell's Gate Ridge came into sight, and the slope gentled under his boots. He was past the worst of the ascent and less than ten yards, ten long strides, from the top of the ridge. The heat grew unendurable. He put up his hands to protect his face.

This is it, he told himself.

Screaming in agony, he threw himself forward in a last, instinctive effort to escape the heat. At that moment, while he was in midair, a wave of superheated gases swept over him. The blast tore off his hard hat and seared his head, neck and hands. It penetrated his shirt and pants where they pressed against his body, burning a partial outline around his backpack.

The hot gases curled in front of his gaping mouth, but his screams kept them out of his throat. The combined effect of his lunge plus the blast of heat threw him face first into the dirt.

The heat wave must have glanced off the slope, because it no longer surrounded him. He could breathe. He regained his feet and tried to run, but waddled forward instead. The straps on his pack had melted, and the pack had slipped down around his legs, held on by the waist belt. As he unsnapped the buckle, he caught sight of his hands. The skin hung in shreds.

He was free, though, free to run. He crossed the ridgetop and launched himself into the East Canyon; if there had been a cliff there, he thought, he gladly would have dived off it. He bounded down the other side in giant steps until he saw Erickson and Haugh. He slowed to a halt.

He asked where everybody else was. The three firefighters looked back at the ridgetop together, but no one had followed Hipke. Spot fires were springing up around them. They had to make an instant decision to deploy shelters or to continue down the East Canyon.

"Is there a safe spot to hang out here?" Erickson asked Haugh, who had been up the canyon twice in the past two days. "No," Haugh replied with conviction.

"Are there any safe areas down the canyon?" Erickson asked.

"Yes," Haugh said, with equal conviction.

Erickson and Haugh put the injured Hipke between them and set off.

ELEVEN

FOR TWO DAYS the top of Hell's Gate Ridge had been a place to camp out, store gear, land a helicopter and construct a fire line as wide as all outdoors. The ridgetop had an eagle's view of the Colorado River Valley down to the notch in the Grand Hogback, where the river and I-70 exited the Rocky Mountains; in the opposite direction it overlooked the maze of gullies that made up the East Canyon. It was about to become a closing trap, a place of white flashes, snowing ash and screaming firefighters.

After Bryan Scholz, the Prineville foreman, shouted his one-word refusal—"No!"—to send more hotshots down the west-flank line, he waited for a response. The decision belonged to the Prineville superintendent, Tom Shepard, but Scholz had made up his mind not to lead anyone down that fire line, no matter what Shepard said.

Shepard let Scholz's refusal stand: There would be no confrontation between superintendent and foreman.

The fire made brisk runs on the slopes of the western drainage. One of Scholz's nine hotshots, Bill Baker, was clearing away branches when "a silence or a calm" made him look up. The fire "crept up on me and then backed off," Baker said later. The flames, to his relief, headed off in the direction of the Colorado River. But that put them on a path toward Michelle Ryerson and her group.

The fire crackled to life directly below Ryerson at approximately 3:50 P.M. Flames the width of a small river licked onto the ridgetop. Ryerson decided to call for the helicopter and its water bucket, but one of her crewmen, Jim Byers, told her to check first with Don Mackey, who might be using the helicopter himself. Ryerson relied for advice on Byers and Loren Paulson, more experienced firefighters than herself; she radioed Mackey, who told her to take the helicopter and its bucket.

Ryerson's flare-up could be fought to a standstill, it could spill over the ridge and die out, or it could overwhelm the firefighters and sweep into the East Canyon, setting it aflame.

Scholz called his nine hotshots around him. "We're going to help our good friends from the BLM," Scholz said, meaning Ryerson and her group, "but if this thing goes to hell, I'll holler, and you come back toward me."

The nine hotshots were not much alarmed. They had arrived on the ridgetop less than an hour earlier and, unlike Scholz, had not been down the west-flank line. Reassuringly, both Scholz and Shepard were with them. The wind had grown stronger, but the fire looked little different to them from the smoky, flaring "twenty-five-acre nothing" they had seen at noontime.

As the nine hotshots went to help Ryerson, a few straggled and they became spread out. One, Mike Simmons, sat down to sharpen a chain saw. The hotshots watched from a distance as Dick Good and 93 Romeo made the first water drop on the flare-up. Only a handful had joined Ryerson by the time the helicopter returned for a second drop.

Radio traffic picked up. The first warning of the blowup by Tony Petrilli, from a post near Lunch Spot Ridge, went unheard by the firefighters along the ridgetop. Petrilli had used the smoke jumpers' radio frequency to alert Mackey—it is common practice for jumpers, pilots, fire-engine crews and others to use their own radio frequencies while having one common "work" channel. But when Jon Kelso, the Prineville squad boss, radioed Shepard that he could see fire "across the draw" from his position on the west-flank line, it had a galvanizing effect. Scholz, overhearing Kelso, told the hotshots who remained near him, "Things are getting complicated."

The smoke column provided its own unmistakable alarm: It turned dense as rope and changed from white to gray-black. The flames along the upper slopes of Hell's Gate Ridge sent up a thickening blanket of white smoke. Winds up to fifty miles an hour whipped the ridgetop. The humidity plunged toward a crackling-dry 8 percent, the lowest level in days.

Everyone on the ridgetop sensed an elemental change. One hotshot, Kip Gray, looked at his watch and announced the time: 4:04 P.M.

"Remember this time," said the hotshot foreman, Bryan Scholz, and others passed along the word: "Remember this time. Remember this time."

At the moment Kelso reported the blowup, Butch Blanco was hauling drinking water to the top of the west-flank line. Until then, the incident commander had felt comfortable with the fire. When Mackey had called a few minutes earlier to renew his request for water, he had sounded no alarm. Blanco knew that Mackey was keeping a watch, because he had overheard him during the afternoon asking Dick Good in 93 Romeo what the fire looked like from the air.

Then, as Blanco said later, "It just happened." The fire blew up, and Blanco grabbed his two-way radio.

"It's going up! Everybody get out!" Blanco said.

Blanco radioed Mackey and told him, "Things don't look good. They're heating up." He also radioed Ryerson and told her to get her crew started for the good black at the first helispot, H1, about 150 yards away in the direction of the Colorado River, the same place Mackey had directed the jumpers on Lunch Spot Ridge to go.

The order baffled Ryerson: What was Blanco thinking? The flare-up threatened to block the path to H1. The wind had turned so blustery that it had knocked off Byers's hard hat a moment before.

The firefighters around Ryerson, hearing Blanco's order,

shouted, "No way!" Blanco must have meant the second helis-pot, H2, about 250 yards in the opposite direction.

"Hey, Butch, I need to reconfirm with you," Ryerson radioed Blanco. Which helispot did he mean, H1 or H2?

Blanco, always the coolest person on a fire, shouted back, "Let's go!" That was enough for Ryerson, and she passed on the order: Head for H1.

"There was chaos; we really had no idea what to do," Ryerson told a reporter afterward. "I was running and scared to death, just trying to keep my composure and pass along what I was hearing on the radio."

By now Shepard had come down from his perch above H2. He had considered the fire "normal" until Kelso's warning. Moments before that, he had been taking photographs of 93 Romeo's water drops on the ridgetop, but he would forget having done so until, many weeks later, the film was discovered by chance in a storage box in Prineville.

Shepard radioed Scholz and repeated Blanco's order: "Go up to H-One."

"Are you sure?" Scholz responded, while passing on the order. H1 was a fine safety zone, Scholz thought, but only if you could get there.

Radio traffic became nonstop. "Are we going to H-One? . . . We're going to do it. There's good black at H-One. . . . Go, go! Get to the safety zone."

A Prineville sawyer, Alex Robertson, and his two swampers, Kim Valentine and Tom Rambo, started to run toward H1; the other hotshots and Ryerson's crew followed in a straggly line. Robertson could hear shouts of "Get up the hill" through his earplugs. Ahead of him Valentine ran wildly, considering the treacherous footing.

Within a few dozen yards an outcropping of rock blocked everyone's way. Valentine whipped around and hurtled back toward Robertson, screaming, "Get down the hill! Get down the hill!" as though she had seen something frightful.

Everyone else bunched up at the foot of the outcropping, unsure what to do next. Sparks and ash showered down on them. A few started to climb into the rocks.

A tip of flame showed itself ahead, rising in a scarlet-orange wave from the western side of Hell's Gate Ridge. The wave began to sweep toward them.

"What the fuck are we doing here?" yelled Simmons, the Prineville Hot Shot who moments before had been sharpening his chain saw.

Scholz, bringing up the rear, heard cursing and saw the wave of flame. "Reverse and move," Scholz yelled, the firefighter's bugle call for retreat.

Valentine burst out of the group and came staggering toward Scholz as though something had sucked the life out of her. Scholz let her pass so he could keep an eye on her as they headed back.

"We can't make it," Ryerson radioed to Blanco. "We have to turn around; we're running back."

As the retreat began, one firefighter, Paulson, cast a backward glance and saw three yellow fire shirts high up in the rock outcropping. They must have missed the order to reverse, Paulson thought, and be unable to see the wave of flame.

Paulson leaned around a rock. "Back down! Back down!" he shouted. The closest of the three shirts hesitated, looked around for the sound and then hurriedly tapped the yellow shirt in front of him, who tapped the shirt farthest along. The three came scrambling down.

The firefighters raced back along the top of Hell's Gate Ridge, surrounded on three sides by fire and gulches. On one side the western drainage was aflame. On the opposite side the East Canyon commenced with a sheer drop-off. There was no turning back toward the rock outcropping—the wave of flame pursued them from that direction. If the fire had crossed the ridgetop behind them while they had been heading toward H1, the trap would be complete.

They neared mental as well as physical limits. The fire had fooled them. Orders were questioned or countermanded as soon as they were given. Those with loose chin straps felt the wind levitate hard hats inches off their heads. Shovels and Pulaskis turned to lead in their hands, and the teeth of chain saws dug into their shoulders.

They approached a black smudge where the flare-up had

crossed the ridgetop but then died out, probably squelched by 93 Romeo's water drops. They scurried past the smudge.

They now came upon a bizarre sight, a firefighter who had stationed himself, Pulaski held as though it were a staff, at the junction with the west-flank fire line. None of them knew Sunny Archuleta by name, but many took fresh hope as he stood his ground and waved them on.

The worst was yet to come.

Shock waves buffeted the firefighters as abandoned gasoline containers melted and gas vapors exploded in flashes "like lightning." They dropped tools, shed packs, stumbled, fell and were helped to their feet. The heels of Baker's boots were separating from the leather uppers, and he literally "walked out" of the boots with each step.

"We're going to try and outrun this thing?" someone called in a voice of disbelief.

Valentine collapsed in a heap, crying hysterically. "It was snowing ash, and there were spot fires all around, and I lost hope," she said later. "I think a lot of us did."

Scholz caught up to her.

"Goddamn it, you're a wife!" Scholz shouted. "You've got a reason to live." He slapped her hard hat. "Get mad, get mad! Get your pack off! You can make it!"

"Go! Leave me here," Valentine said.

The wave of flame swept toward them, and Scholz felt the beat of his heart rise from his chest to his throat; the expression having your "heart in your mouth," he discovered, could be a literal description of fear. The flames around him were "a nasty orange, a malignant orange, not a friendly clown orange, just a deadly orange base with a black wall on top of it—and that freight-train roar." He yearned to put on blinders and follow the heels of someone else's boots, but he had to say something to Valentine.

"I would never leave you," Scholz said, and later that would be the one thing she would recall out of everything he told her.

"Come on, come on, we've got to move!" Scholz said. Valentine wriggled free of her pack and fought to her feet, wobbly but upright. The drop into the East Canyon was steep but no longer sheer; she and Scholz headed over the side with Scholz yelling to others, "Come on Prineville, come on you BLMers!"

A tall figure—Robertson, the Prineville sawyer—came pounding out of the smoke behind them, high-stepping to avoid stobs. Robertson never looked back and so had no picture of himself riding a flaming wave, as others did, but he felt heat rising on his back and side.

Robertson pulled out his earplugs to gauge the fire's strength and heard an ungodly roar. He launched himself over the lip of the ridge in headlong flight, his boots skidding on loose shale. He bounded past the carcass of a long-dead animal, big enough to be a mule deer or elk.

Ryerson watched others descend into the East Canyon. Embers sizzled on her shirt. This is a run for your life, she told herself, but no place looked safe. The East Canyon was a daunting sight, endless ravines that already might be in flames. Instinct and training told Ryerson to "stay high," out of the gulches.

She reached for the fire shelter in her backpack, fumbling blindly, and someone grabbed her hand. It was Paulson, her ex-boyfriend, and she felt a rush of the old intimacy.

"Are we going to die?" she asked.

"I don't think so," Paulson replied, trying to calm himself as well as her. A few steps away Valentine was rising to her feet. "She's making it," Paulson told Ryerson. "You can make it, too. You're in good shape, you work out, you'll be fine."

Paulson and Ryerson made a quick pact: They would stick together whether they deployed fire shelters or made a run for it. They helped each other take out shelters and started off, holding the shelters at the ready. Within a few steps they hooked around a small knoll and found momentary protection. Another firefighter, Archuleta, appeared out of the smoke. After waiting too long to retreat, Archuleta had managed to keep a step ahead of the wave of flame. Ryerson asked him which way they should go.

"How the hell should I know?" Archuleta replied.

"Screw it, I'm deploying," Ryerson told Paulson and started to shake out her fire shelter. Afterward investigators concluded that she and Paulson might well have survived by deploying shelters in this particular spot.

Paulson, though, didn't like the look of the place. The fire was crossing the ridge ahead of and behind them; in a minute the brush around them would catch fire. Fire shelters can withstand hours of radiant heat, but no more than seconds of direct flames.

"Do you want me to leave you here? You deploy, and I'll go downhill," Paulson said, trying to shock Ryerson into moving. The tactic worked too well; Ryerson looked stunned—they had agreed to stick together!

"I'm not going to leave you," Paulson said, relenting, "but get your ass downhill." They quickly caught up to Scholz and Valentine, and the four of them, with Ryerson leading, headed down the East Canyon.

Archuleta, coming along behind them, picked up Valentine's pack and slung it over his shoulder; it had a familiar look. Prineville Hot Shots, it turned out, carried fire packs made in the off-season by Missoula smoke jumpers, and Archuleta's job was to sew them together. Others shouted at him to drop the pack, but he carried it with him every step of the way.

Valentine kept falling in a heap, legs skewed at nearly impossible angles. When Paulson handed her his water bottle, warning her to keep moving while she drank, Valentine turned icy calm.

"Oh, you're so nice," she said with drawing-room politeness. She's really lost it, Paulson said to himself.

As the retreat progressed, Blanco made his way to the clearing at H2 for a better view of the fire. He radioed the BLM office in Grand Junction. He had a blowup on his hands, he told the chief dispatcher, Flint Cheney, and it threatened homes in Canyon Creek Estates and along I-70. He needed air tankers with retardant right away for home protection. Cheney asked what good tankers would do, considering the terrain.

Blanco replied that they had to try something. He said nothing about the threat to the firefighters, but with flames seconds away and the BLM office ninety miles distant, there was little to say.

And now, after more than three days of delays, aborted action and denied or half-filled requests, the supply floodgates for the

South Canyon fire swung wide open. At 4:08 P.M. the Western Slope Coordination Center, contacted by Cheney, issued a multiple order: one air attack plane to direct operations, a second helicopter, one lead or scout plane and two air tankers, all to be dispatched "ASAP," as soon as possible.

Cheney wrote in the log at 4:11 P.M.: "Blanco: Losing fire on side closest to homes. Needs retardant."

Blanco called again, two minutes after Cheney made that entry, and said that the fire had become an immediate threat to Canyon Creek Estates. He asked that fire engines be sent from Glenwood Springs to defend homes. Seven minutes after that, at 4:20 P.M., the log records a last appeal from Blanco: "Call Garfield County to begin evacuations."

Blanco made one further request after matters had become so chaotic that Cheney had stopped keeping the log. Blanco told Cheney that there were "probably going to be some burns" and asked for the medical-evacuation helicopter from St. Mary's Hospital in Grand Junction.

Shepard meanwhile had joined Blanco at H2, and both of them began waving people into the East Canyon. "Go down, go down!" they shouted.

"Look out in front!" the firefighters yelled. "Don't lead us into fire! Get your shelters ready." As they stepped off the ridgetop, they met an erratic wind, strong in one place, absent in another, blowing up the East Canyon. That wind had one marvelous initial effect: It checked the prevailing west wind at the top of Hell's Gate Ridge and, for vital minutes, kept the main fire out of the East Canyon. But it also created a new danger. If small fires started in the East Canyon, this counterwind would fan them into another inferno.

That began to happen: One of the Prineville Hot Shots, Tom Rambo, saw flames starting up below him in the East Canyon. Remaining on the top of Hell's Gate Ridge would be like standing on the sun, Rambo told himself. He shouted to a fellow hotshot, Tony Johnson, "We've got fire below us! Find a place to deploy."

Rambo and Johnson plunged off the ridgetop into the canyon together; Johnson was thinking what a good story this was going to make to tell his brother, Rob, when they saw each other again. Rob was with the Prineville Hot Shots on the west-flank line.

Rambo and Johnson hurtled past the spot fire, never taking out their shelters.

As Blanco and Shepard prepared to leave the ridgetop, they caught sight of a stack of red gear bags in the brush. They kicked the bags into an open space to protect them. A hopeless gesture if there ever was one, Shepard thought later. One of the helicopter crewmen lent a hand—Shepard was never sure whether it was Tyler or Browning.

Shepard next saw Tyler and Browning jogging along Hell's Gate Ridge carrying their heavy flight helmets, as though following a plan. It would have been Tyler's plan as foreman and local expert, and he of all people would have put safety first, remembering his lost crew. Shepard and Blanco yelled at the pair, "Come on, come on! Go down, go down!" pointing toward the East Canyon.

"Run the ridge, run the ridge," Tyler and Browning shouted back. If they followed their present course, they would reach the junction of Hell's Gate Ridge and the peak of Storm King in about three hundred yards, then face a choice.

They could circle around on high ground above the East Canyon, which would be safer in theory than descending into the canyon itself. Or they could head in the opposite direction, across the top of the western drainage, and try to beat the blowup heading that way. Either choice would keep them on high ground for a helicopter pickup.

One hotshot, Louie Navarro, glimpsed the two men grabbing handfuls of brush to pull themselves along. Navarro yelled at them to go into the East Canyon, but they were too far away to hear. In the next instant Tyler and Browning disappeared behind intervening flames. Nearly forty-eight hours would pass, nearly two full days, before their fate became known.

The reflector strips on Navarro's hard hat bubbled from radiant heat. He felt a pain along his shoulders and thought it was neck strain, until he realized he had been lightly scorched. The fire "gasped for air and then surged like a tidal wave." Trying to deploy a fire shelter, Navarro told himself, would be like trying to fold sheets in a hurricane. He half dove into the East Canyon and was met by an up-gulch wind strong enough to stand him up.

As the flames broke across the ridgetop, they ignited a box of fusees. The explosion helped drive Shepard and Blanco the last few feet into the East Canyon. There they became separated. Shepard found himself surrounded by an eerie calm, but he took no comfort from it. His thoughts were elsewhere, with the nine Prineville Hot Shots left behind on the west-flank line, "those nine folks that I felt I had lost—that's where I was at."

SCHOLZ AND VALENTINE, Ryerson and Paulson, Navarro, Rambo, Robertson and the others had become a battered army wishing for a quick exit from the scene of a rout, but this was not to be. They had broken into twos and threes in the rush off the ridgetop; now they tried to reestablish contact by yelling, calling out names of people they had seen and asking who was missing.

The flame wave had spent itself on the ridgetop. The fire now rose in a solid wall above the ridge as high as flames can reach, about 150 feet. The pause in the fire's advance gave the firefighters a priceless head start down the East Canyon, but they had to hurry.

The three men who had escaped from the west-flank line— Kevin Erickson, Brad Haugh and Eric Hipke—walked together down the canyon until the vegetation thinned and they felt secure enough to make a brief halt. Erickson and Haugh poured water over Hipke's arms and hands to cool the burns and wrapped wet strips of T-shirt around his hands.

As they resumed the march, Hipke held his arms out in front of him and chanted "Oh, man; oh, man." Erickson and Haugh worried that he was heading into shock. They asked every few steps, "What's your name? What day is it?" until Hipke replied in exasperation, "Wednesday—let's get the hell out of here."

No Prineville Hot Shot, from the rank and file to Shepard and Scholz, had been in the East Canyon before or had scouted it from the air, and they had no faith that it led to safety. "This is bullshit, this is wrong," one hotshot shouted. Several later described the situation at this moment in the same terms: "total, organized panic."

They called out, "Shelters ready! Shelters ready!" until most had shelters in hand. "Who do you have with you? Who have you got?" Several of them passed Tom Shepard and tried to exchange a word with him, but he barely responded.

Butch Blanco went from group to group, urging them down the canyon. "Who are you?" one of the Prineville Hot Shots challenged. "I'm Blanco," he replied.

One hotshot, Baker, stumbling as his boots fell to pieces, knew he ought to stop and deploy his fire shelter. He reprimanded himself with gallows humor: If the fire spots in front of us, and the wind is blowing up the canyon at fifty miles an hour, then there's going to be no time to deploy . . . and I'm going to feel like a real ass just before I die.

The brush knocked off Baker's hard hat, and turning to retrieve it, he found himself in the path of the fire, slowly descending into the East Canyon. Oh, God, Baker said to himself, I need the hat. He reached out and snatched it back.

Several small groups banded together and crawled onto a broad bench, a flat area surrounded by deep gulches. From there they scanned back along the ridgetop. "Where's the rest of the people?" one asked, meaning those from the west-flank line—Jon Kelso, Tami Bickett and the others. They milled about, hunting for someone with a radio.

Jim Byers, the BLM veteran, climbed the bench after them. "We can't be up here. We have to leave," he told them as diplomatically as possible.

The elevated bench seemed safe to the hotshots: Why move? Their friends, half of the Prineville Hot Shots, were somewhere back there, perhaps even now following them over Hell's Gate Ridge. They resisted a few moments longer, then began to leave, once again in twos and threes.

The retreat became more orderly as everyone settled to the task. The East Canyon narrowed into a single, twisting gorge, driving them together. Sarah Doehring and Derek Brixey joined up again and found themselves behind a slow-moving group of four—Michelle Ryerson, Loren Paulson, Bryan Scholz and Kim

Valentine. Valentine continued to stumble and require assistance. Paulson tried to lift everyone's spirits with a joke: "This is why we get hazard pay."

Doehring offered to help with Valentine, but there was nothing for her to do. They don't need our help, Doehring told herself, so why should we walk slowly? She and Brixey went around the four and continued on at a swift pace.

The gorge began to twist and turn. Near-vertical sides closed off the view to the front and rear, creating the effect of a box canyon advancing in lockstep with the crew, its conclusion nightmarishly beyond reach. Then a sliver of blue sky and sunlight flashed ahead—the opening of Hell's Gate, and beyond that I-70 and the Colorado River. "Suddenly there was the sun, and I was looking back over my shoulder at the smoke column, and I knew we were going to make it," Scholz said.

Smoke vented out Hell's Gate. Blanco stopped short of the canyon's mouth, and as everyone passed by, he began compiling a list of names.

Hipke, Haugh and Erickson were among the first to exit the canyon; the others filed out Hell's Gate behind them—Doehring and Brixey, Ryerson and Paulson, Scholz and Valentine. By then Hipke was in shock. He stretched out on the ground, and someone wrapped him in a space blanket. He could see Prineville blue hard hats gather above him, the dirty, ash-streaked faces unrecognizable beneath them. The only Prineville Hot Shots Hipke knew about were those who had been with him on the west-flank line. They must have made it out, he thought.

A ragged line formed, waiting to empty canteens and water bottles onto Hipke's head, hands and arms. Louie Navarro had filled his canteen with a special high-caffeine coffee to counter fire-line fatigue. What the hell, Navarro said to himself, and poured the coffee over Hipke. It seemed to give relief.

The burns on Erickson's arms had begun to throb. He paced back and forth along the side of the highway, cursing to himself and muttering, "I don't believe it; I don't believe it." Sarah Doehring, among the first to exit the canyon, rushed over and gave Erickson a big hug, realizing only when he recoiled that he had been burned.

The storms that caused the fire on Storm King and the blowup came from the west, the same perspective as shown in this photo. Canyon Creek Estates is in the foreground; Hell's Gate Ridge is the heavy ridge in the upper right-hand corner.

Mann Gulch and the Missouri River in west-central Montana at the time of the Mann Gulch fire in 1949.

Storm King Mountain from a similar perspective. Note the similarities to Mann Gulch: the funnel-like gulches, the broad river—here the Colorado—and the heavy peak in the background. In both places, fires, fanned by winds from the rivers, exploded in the gulches.

Don Mackey, the "heart and soul of the smoke jumpers," in jump gear.

The Prineville Hot Shots in their only group photo, taken at Crater Lake in Oregon on July 4, two days before the blowup on Storm King. The Hot Shots had finished work on a fire in Oregon and thought they were headed to Prineville. A tourist took the photo with one of their cameras.

(Rear from left) Kip Gray, Rob Johnson, Scott Blecha, Alex Robertson, John Kelso, Bonnie Holtby, Mike Simmons.

(Center from left) Tom Shepard, Kathi Beck, Tony Johnson, Terri Hagen, Tami Bickett, Levi Brinkley, Kim Valentine, Bryan Scholz.

(Front from left) Brian Lee, Louie Navarro, Tom Rambo, Bill Baker, Doug Dunbar.

The Prineville Hot Shots form a human chain to unload their gear from a school bus at Canyon Creek Estates.

The fire smolders on Hell's Gate Ridge as Hot Shots unload gear.

The Hot Shots ready fire packs before flying by helicopter to Storm King Mountain.

The fire burning on Hell's Gate Ridge on July 4, about thirty-six hours after it ignited, seen from Canyon Creek Estates. No one yet has fought it.

Smoke jumpers use chain saws and hand tools to begin the west flank fire line the morning of July 6, the day of the blowup. Note the dense Gambel oak.

Wayne Williams was among the smoke jumpers who arrived by bus at Canyon Creek Estates just as the fire blew up at 4:00 P.M. on July 6. He took this photo of the blowup moments later.

Smoke jumpers bused to Canyon Creek Estates watch helplessly as the blowup progresses.

These photos, arranged
chronologically, were taken at
4:11 P.M. on July 6 looking
down toward the west flank
fire line. The first photo includes
three tiny human figures,
difficult to make out except in
close-ups, in the fire line.

Sunny Archuleta took the
first photo; Sarah Doehring
took the last two, showing
Archuleta gazing spellbound
at the blowup.

After the bodies were removed from the west flank fire line, yellow fire shirts were put down to identify where each had been found.

The steepest portion of the west flank line was a fifty-five-degree incline.

The fire melted survivor Eric Hipke's pack and video camera.

Wooden crosses at first marked where fire-fighters died: Jim Thrash's cross is to the right, Kathi Beck's to the left.

A squirrel caught in midflight as it raced the blowup.

DAN JACKSON

The fire left this pine tree at the top of the west flank fire line in the shape of a cross. By the next spring, the branches had broken off and the tree had become a spike.

CASEY CASS

DON MACKEY
1960 – 1994

Nadine Mackey plants tulips near Don Mackey's and Bonnie Holtby's newly installed granite crosses during a break in the stormy weather of April 1995. The eagle feather and trap at Don's cross were left as tokens by friends.

An ambulance showed up within minutes. Attendants loaded Hipke onto a gurney. Erickson climbed in beside him. The ambulance drove them to the Valley View Hospital in Glenwood Springs, where attendants began cutting off their shirts.

People clustered around Valentine, who sobbed inconsolably. She cried out over and over, "They didn't make it! They didn't make it!"

"They're fine, they got out," someone told her, but it had no effect.

Anyone with a radio drew a crowd. Calls went out for the rest of the hotshots, smoke jumpers—anyone who would answer. A voice finally responded from the helicopter base at Canyon Creek Estates, but it had no news to offer. Shepard and Scholz learned that there had been fire-shelter deployments on the mountain but couldn't pick up any details.

Then came an electrifying report: "Kelso plus four on the road!" Jon Kelso must have made it out to I-70 with at least three others from the west-flank line. No one knew who started the report, but it spread through the crew in waves, dying out and then reigniting on a fresh surge of hope: "Did you hear it? Kelso plus four, Kelso plus four."

It was time to leave Hell's Gate. The group split up, boarding trucks that had pulled up in the meantime. The smoke jumpers and supervisors headed for Canyon Creek Estates, the hotshots and the rest of the BLM–Forest Service crew for the BLM office in West Glenwood. "Think good thoughts," Scholz told the hotshots as they left.

The retreat down the East Canyon had taken more than an hour, but everyone who had been on the top of Hell's Gate Ridge when the blowup occurred was now accounted for, with the exception of Tyler and Browning. Miraculously, no one had broken a leg or sustained other serious injury in the descent. Blanco was the last to leave the canyon, emerging at about 5:30 P.M. Within a half hour, by 6:00 P.M., the fire had gutted the East Canyon from the top of Hell's Gate Ridge to the highway.

When the trucks arrived at the BLM office in West Glenwood, some of the BLM crew members began to celebrate. "Goddamn, we made it!" they shouted, slapping hands in high fives.

The Prineville Hot Shots were subdued. With Shepard and Scholz off trying to find out what had happened, there were nine hotshots left—Bill Baker, Kip Gray, Tony Johnson, Brian Lee, Louie Navarro, Tom Rambo, Alex Robertson, Mike Simmons and Kim Valentine. They gathered separately at a picnic table in the storage yard at the rear of the BLM office. Repeated attempts to raise Kelso by radio had failed, and "Kelso plus four" had lost its magic.

A smoky haze collected around the storage yard. Nearby, homeowners began spraying roofs with garden hoses. Sirens shrieked from passing fire trucks, police vehicles and ambulances. On Hell's Gate Ridge, visible from the picnic table, the wall of flame soared into the air.

Robertson, the Prineville sawyer, broke the silence. "I'm not a religious man," he said, "but I think we should pray." The hotshots bowed their heads and began to mumble half-remembered fragments of the Lord's Prayer. Then Valentine, speaking for them all, recited a complete prayer.

"After that, I pretty much went black; I don't remember anything," Valentine later said in an interview with her college-alumni magazine, her only public account of her experiences. The hotshots found a cot for Valentine to lie on until an ambulance arrived. She, like Hipke and Erickson, was taken to Valley View Hospital.

Blanco in the meantime had gone to Canyon Creek Estates to compare his list of names with other lists. There were fourteen firefighters missing: the two helitacks, Tyler and Browning; and twelve people from the west-flank line, including Don Mackey and two other smoke jumpers—Jim Thrash and Roger Roth—the hotshot squad bosses—Tami Bickett and Jon Kelso—and seven squad members. Tyler and Browning had been alive and running when last seen; the twelve had been on the west-flank line when flames engulfed it.

Blanco came to the BLM office in West Glenwood and wandered through the storage yard, appearing stunned. Someone asked him how he was doing. "I just killed twelve kids," Blanco said, a statement he would make in different forms many times in the future.

The fire began to probe the hills behind the BLM office, and a decision was made to evacuate the place. The firefighters climbed back into the trucks—everyone except Haugh, who decided to stick it out at the BLM office. Haugh hid in a storage room until everyone left and then, looking for something to do, found a broom and began to sweep the floor.

The trucks headed for Two Rivers Park, near the center of Glenwood Springs on the bank of the Colorado River at its junction with the Roaring Fork River. The rolling lawns were a lush, welcoming green. A softball game was in progress. Music flowed from a band shell, part of the Glenwood Springs Summer Jazz concert series. A festive crowd sported clean clothing in brilliant whites and sunny pastels. There were coolers of iced sodas and hampers brimming with sandwiches and bags of chips. Children and dogs played around picnic tables.

The trucks drew up in the parking lot, and the firefighters slowly climbed out, their yellow shirts and green pants filthy, their faces gray with ash and exhaustion. They slumped onto the grass. A few people broke away from the crowd, approached and asked a hesitant question or two. Soon a widening stream of concert-goers formed a link between the coolers and the firefighters, offering sodas and sympathy.

A minivan pulled into the parking lot and Shepard and Scholz stepped out along with two people in civilian clothes. They introduced the civilians, a man and a woman, as crisis-intervention personnel. Everyone would be going to a motel, Shepard and Scholz said, to start sorting things out.

TWELVE

THE SOUTH CANYON fire spread its effects beyond Storm King Mountain to homes in the foothills, vehicles on I-70 and fire offices across Colorado, and eventually to more distant places and into uncounted lives. It became a spectacle for some, a tragedy for others. It would give rise to rage and blame, compassion and forgiveness; it would destroy some careers, enhance others and eventually work changes in the way fire is fought. Like the Mann Gulch fire, it would burn in the lives of those it touched long after the last ember died out.

The top regional fire supervisors, Pete Blume and Winslow Robertson, heard Blanco's report of the blowup when it came over the radio shortly after 4:00 P.M. in the BLM dispatch office in Grand Junction. Until then everything had seemed "just fine" to Robertson. Blanco had called in less than an hour earlier and mentioned nothing disturbing except an increase in winds, common in late afternoon.

True, one delay had followed another in rounding up reinforcements for the fire. But the bus with twenty-eight fresh smoke jumpers should be arriving at Canyon Creek Estates about this time. That would increase the number of the crew to nearly eighty, surely enough to handle a well-behaved fire. In normal times Blume as fire-management officer would have driven to a fire this size to look it over for himself—in past years he had visited every fire of more than ten acres in his district. Blume had made those trips, he later said, to "get a sense of exactly what the fire is like, the aspect and steepness of the slope and the fuels, what problems the IC is having."

In this busy fire season, however, Blume remained at the administrative hub in Grand Junction, relying on his assistant, Robertson, to be his eyes and ears in the field. This afternoon, though, Robertson, too, was in the dispatch office, a circumstance that was never explained.

The blowup came as a complete surprise to them. "With all that we had going on, it was one of the last places I expected anything like this to happen because of the level of experience and the number [of firefighters]," Blume later told fire investigators. "Between the Type I hand crews, and all the jumpers are Type I, and I had an experienced engine foreman—"

"But now Mann Gulch, wasn't that with Type I's?" the questioner interjected.

"It was smoke jumpers," Blume said, "so yes, it was."

When Robertson heard Blanco warn of a threat to homes, he hurried from the dispatch office, no longer the reluctant warrior. "I jumped in my truck and headed up there," Robertson recounted to fire investigators. "And then after that I was incident commander of the fire."

The bus with the twenty-eight smoke jumpers, meanwhile, drew in sight of Storm King. George Steele and his seat mate, Bob Hurley, had been warned that the fire would be a big one, but as they approached, they remarked to each other how little smoke they could see. The wind must be holding the smoke to the ground. Steele and Hurley had been released from the Wake fire that morning, after spending the night on a nearby fairground. They had arrived groggy and tired in Grand Junction at about 9:00 A.M., but had stepped forward when the jump coordinator asked for volunteers for a "ground pounder" that would require hiking: the South Canyon fire.

In the few minutes it took for the jumpers' bus to pull off I-70 and make its way to the meadow at Canyon Creek Estates, the smoke column rose up and became gigantic. The twenty-eight jumpers disembarked with a helpless feeling. The fire had gone from being invisible one minute to overwhelming the next. The wind, howling at forty to fifty miles an hour, peeled smoke off Lunch Spot Ridge; the jumpers caught momentary glimpses of mirrorlike dots.

"There's shelters deployed!" one jumper shouted.

"Oh, my God," another said. For a long time none of them said anything more.

At about the same time Bruce Meland, an electrical engineer, was driving past Storm King Mountain on his way home to Bend, Oregon, twenty-five miles from Prineville. Meland had been in Denver covering a story for a magazine, *Electrical Times*, which he published.

Meland pulled to the side of the road and took out a camera. He had written a master's-degree thesis on grass fires and at first thought the blowup was an example of his academic specialty. He quickly realized, however, that the smoke was too heavy for burning grass.

"It looked like Mount Saint Helens going off," Meland said, recalling the massive volcanic eruption on May 18, 1980, in Washington State. Meland wondered why no support airplanes or helicopters were in sight, considering that the fire was so far advanced. He took a series of photographs and was home in Bend before he learned about the connection to the neighboring town of Prineville.

IN GLENWOOD SPRINGS, Karen Olsen, a Chicago suburbanite, was relaxing with her family in the turn-of-the-century elegance of the Great Swimming Pool when she felt a wind stir. The place could not have been more beautiful; the Victorian bathhouse and pool, fed by the hot springs, had undergone an extensive restoration, completed earlier in the year.

The sky turned black and yellow, and an abrupt wind blew dishes off tables. Olsen knew a storm warning when she saw one—she once had been a professional photographer of sailboat races. The fire came over Storm King and headed for the town. This was no time to be playing in the water, Olsen told herself, and gathered her family as sirens began to wail.

THE TWENTY-EIGHT SMOKE jumpers gathered along the rail fence surrounding the meadow at Canyon Creek Estates and scanned the mountain for signs of life. The fire shelters on Lunch Spot Ridge appeared as inanimate as cocoons.

A news reporter asked one jumper, Calvin Robinson, if anyone was holding the fire lines. "They're running for their lives," said Robinson, who was well acquainted with blowups. His father, Laird, once a Missoula smoke-jumper foreman, had done much research on the Mann Gulch fire, helping to make it known outside the fire world.

The helicopter crew in the meadow kept watch as well. Only a few minutes earlier Rich Tyler had been on the radio with his assistant foreman, Bruce Dissel, telling him to make arrangements for the crew to spend the night in Glenwood Springs; the fire was taking longer than expected to control. Other than that, Tyler had sounded as though everything was going well. Dissel had gone to a nearby home to telephone a motel when one of the crewmen, Steve Little, raised him on the radio.

"I think you better get out here and see this," Little said.

"Can't you handle it? I'm on the phone," Dissel said.

"No, I think you better get out here."

As Dissel stepped outside, he couldn't believe his eyes. "It looked as though an atomic bomb had gone off," he said later. He joined the others watching the thundering smoke column. Someone sighted "tinfoil" flashing from Hell's Gate Ridge. At 4:29 P.M. Dissel reported by radio to the BLM's Grand Junction dispatch office: "Shelters deploying, quite a few."

The wind was so heavy that it held 93 Romeo on the ground. There was nothing anyone in the meadow could do but wait.

In Grand Junction the initial order for multiple aircraft had gone through, and the first of what would become an air fleet began to roll for Storm King. "We tried to scramble anything to get it up there for them," Cheney said.

The first orders to the pilots, based on Blanco's alarms about the threat to homes, stressed the defense of property. By the time a threat to life became an issue, concern for property had become an entrenched mind-set. This started what amounted to a tug-of-war, which no one wanted and several people tried to head off, about which priority came first, lives or property. The conflict lasted for the rest of the sixth and into the next two days, July 7 and 8. It grew more complex as other motivations came into play, from rules about who was supposed to search for survivors to the desire by fire managers to show a watching political hierarchy—including the governor of Colorado and eventually the chiefs of the BLM and the Forest Service, the secretaries of interior and agriculture and the President of the United States—who was boss in the woods.

As with the mixup that resulted in the wrongful name "South Canyon fire," an early mistake, in this case the focus on the threat to property had a later effect no one intended—and this time the consequences were grave. The overemphasis on the threat to property threatened pilot lives on the sixth and seriously disrupted the search for survivors on that and subsequent days.

By 4:32 P.M. three airplanes—an air-attack plane with an air supervisor, a lead plane and a retardant tanker—reported themselves rolling or airborne for Storm King. Another tanker and three more helicopters would follow. It was a quick air hop, fifteen to twenty minutes, from Walker Field to the fire.

The air-attack supervisor, Dick Ferneau, a thirty-year Forest Service veteran, had already made the trip once that day, flying past Storm King at noon to determine if a tanker "air show" would be useful. The fire had appeared "calm, quiet, pretty darn good," and the ravines too narrow for the bulky tankers. With afternoon winds predicted on top of that, Ferneau had recommended against the air show he now found himself leading.

When Ferneau arrived in late afternoon for the second time, the mountain was hidden behind a smoke column more than forty thousand feet high. The wind had grown fierce and erratic. Fire descended into the East Canyon.

Ferneau radioed Grand Junction at 4:42 P.M. and reported that it was too smoky to conduct a major air-tanker operation: "Structure protection a nil, smoke bad," the log reads. The airplanes flew in a stack with Ferneau's the highest, at about eleven thousand feet, giving him an overall view. The smaller lead plane flew low to scout; the tanker orbited in a holding pattern to the west of Storm King.

The pilot of the lead plane, Cliff NaVeaux, an ex–smoke jumper who had lost a leg in a skiing accident, braced one hand against the roof of his cockpit and dipped his light craft into the V of the western drainage. As the plane twisted and turned, boxes of spare parts flew out of the rear seat, landed in the copilot's seat, and bounced back again.

"I remember thinking, Why do they always wait so long for air support?" NaVeaux said later. As he flew above the drainage, a metallic glimmer winked up at him. NaVeaux dropped lower.

NaVeaux steadied the plane and radioed to Ferneau that fire shelters had been deployed. The two airmen, worried by the stillness of the metal tents, spun radio dials trying various frequencies to raise survivors. The impression of lifelessness, however, was misleading.

Inside the fire shelters Tony Petrilli and the others could shout to each other, peek out and squirm around on elbows and knees. Quentin Rhoades tried a joke: "Hey, anybody got a cigarette?" The fire made several runs past them, once close enough to send firebrands swirling around the shelters. With the sun beating down and radiant heat coming in waves, temperatures under the shelters rose to well over 100 degrees.

"My feet are burning!" cried out Eric Shelton, whose shelter, directly below Petrilli's, had begun to slide into a burning stump. Petrilli and others scooted uphill to make room for him.

If only this were over, Petrilli thought, if only I were home with my sons—Vince, Kyle and Dominic. When he heard Ferneau's voice, he quickly responded: Everyone around him was okay, Petrilli said, "but I don't know what's going on below us."

The patch of Gambel oak near the jumpers remained unburned, an invitation to another fire run. But it was a difficult target for a tanker. The pilot would have to fly in a long banking turn above Hell's Gate Ridge, head into the wind, level out and then release the retardant virtually before seeing the target.

At 4:55 P.M. Ferneau told the BLM's Grand Junction dispatch

office that he intended to "put retardant in with the crews," meaning commence an air-tanker operation. While setting things up, Ferneau asked Dick Good in 93 Romeo to make a reconnaissance flight of the mountain. Conditions had calmed enough by 5:08 P.M. for Good to take off. Ferneau, piecing together reports from Good and Petrilli, told the BLM dispatch office at 5:21 P.M. that he could account for nine smoke jumpers—Petrilli and his group. "Unknown about anyone else," the log reads.

At that same minute, 5:21 P.M., Winslow Robertson reported that he had arrived at Storm King and taken over as incident commander. Robertson told the dispatch office that he had made contact with Blanco and was working up a list of the missing. The fire appeared to be "blowing away from structures," Robertson reported, a relief for everyone. On the way to the fire Robertson had put in a request for two heavy fire engines for home protection and four more ground crews, or eighty people.

There was so much radio chatter about structure protection that a tanker pilot, Mark Bidgood, was startled to find a rescue operation under way when he arrived minutes behind the other airplanes. Bidgood radioed the dispatch office for clarification: What was the number-one priority, structures or people? The dispatcher, faced with a clear choice, replied, "People."

Bidgood drew the assignment to make the retardant drop on Lunch Spot Ridge. He began the long turn over Hell's Gate Ridge, a maneuver captured on videotape and in photographs. The four-engine tanker arced toward its target. Red mud gushed from the underside of the tanker and streaked downward against a backdrop of gray-white smoke. The mud, turning pink as it thinned out, splattered onto the oak patch near the fire shelters.

Petrilli, unable to see that far, called for another drop. A second tanker, T10, flew the same looping pattern and dropped a load of retardant directly on the shelters.

"Right on the money," Petrilli reported. "We're fine now."

The smoke column, meanwhile, had become a magnet, drawing aircraft and vehicles of every sort. Private-plane operators flew overhead for a bit of sight-seeing despite emergency flight restrictions for the area. Fire engines, ambulances, police and news vehicles formed ranks in the meadow and along the frontage road paralleling I-70.

One homeowner toppled trees behind his house with a chain saw. Another turned out two horses, who began to roam the frontage road. The evening rush hour began, clogging I-70. Robertson had tried to order the highway closed but lacked the legal authority; state highway police did maintain one westbound lane for the exclusive use of emergency vehicles.

Communication demands overwhelmed the radio network.

"[There were] turbulent conditions, poor radio communications and no info on the status of the fire crews," the scout pilot, NaVeaux, said later. "A very chaotic time . . . more tankers, more helicopters showing up, jamming every and all radio frequencies."

The pressure to defend structures continued unabated. "We did not know from whom or from where the information was coming," NaVeaux said.

The BLM dispatch office in Grand Junction had lost contact with Robertson and Blanco, and in desperation Cheney telephoned the BLM's office in West Glenwood. Cheney arranged for Bob Elderquin—a "guy with a truck and a radio"—to act as a communications relay. At 6:20 P.M. Elderquin reported that evacuations had begun at Ami's Acres trailer park and at residences along Mitchell Creek Road along the foot of Storm King.

The Garfield County sheriff's office announced by megaphone that residents had fifteen minutes to grab belongings and leave. An orange glow appeared in the sky; clouds of ash blew through the air.

"I knew it was time to go when I could feel the heat and see the flames coming," one resident, Jeff Jensen, told a *Glenwood Post* reporter. "When I came home from work there was so much smoke I had to drive with my lights on. I grabbed the photo albums and insurance papers and put a lawn sprinkler on the roof." Eventually more than two hundred people abandoned their homes.

By 6:20 P.M. the air-attack supervisor, Ferneau, decided that the air show had become too risky. He radioed the BLM dispatch office and said that the air tankers should be released to return to

Grand Junction and the helicopters should be grounded. The dispatcher, apparently acting on orders from Robertson, told him to hold all aircraft at Storm King.

A minute later NaVeaux made a separate call to the Western Slope center, sounding the same alarm. The air had become "too rough" for continued operations over Storm King, NaVeaux said.

Robertson had become increasingly concerned about the threat to homes. At 6:27 P.M. Robertson contacted the Western Slope center and renewed a request for retardant drops for homes near I-70. He added that the fire was burning away from the area; so far "structures good," he reported. (That entry contains the first accurate count in the log of the missing: "9 from crew, 3 jumpers, 2 93R," shorthand for the nine Prineville Hot Shots who had been on the west-flank line, the three smoke jumpers who had been with them and the two helicopter crewmen, Tyler and Browning—fourteen in all.)

The situation headed out of control. Ferneau and Robertson never established direct radio contact—Ferneau said he did not even learn Robertson's name on July 6. Passing messages anonymously through intermediaries guaranteed a lack of discussion and a climate where misunderstanding could flourish. The record of these exchanges is so garbled that it has to be explained by one of the go-betweens, Flint Cheney.

Ferneau radioed the BLM dispatcher at 6:30 P.M. to complain about flying conditions for a second time. Fuel was running low, Ferneau said, and the air tankers were "leaving the area."

The dispatcher replied that a "ground contact," presumably Robertson, insisted that the air tankers "remain on site."

A further exchange was recorded piecemeal: "It is unsafe to drop any retardant, Amy's [sic] Acres," the log reads, presumably Ferneau's judgment. That is followed by a long blank space and then "don't release," presumably Robertson's response to Ferneau.

Cheney, who was in the middle of the exchanges, later filled in some of the blanks. "I remember Winslow asking if those air tankers could get down in Mitchell Creek because that's where the fire was starting to push into West Glenwood. [Ferneau] said the tankers were so low on fuel they had to get back to Grand

Junction. I remember Winslow saying, 'We've got to get some retardant into Mitchell Creek,' and all of a sudden his air tankers are leaving because they're low on fuel. So there was some frustration." The phrase "some frustration" was an understatement, Cheney acknowledged; others remember hearing shouts and yells on the radio.

Ferneau later gave his version of the exchanges to fire investigators: "Some asshole on the ground told me to stay there and orbit. . . . The IC [Robertson] ignored some fifty years of fire and air experience in requiring lead, air attack and tankers to remain over the scene."

Ferneau eventually sent the air tankers back to Grand Junction but ordered two helicopters, better suited to rough terrain, to commence dropping water on homes near I-70. Two other helicopters remained on the ground at Canyon Creek Estates in case medical evacuations became necessary.

The helicopter pilots conducting water drops soon found the sight of the Colorado River irresistible and began flying over rush-hour traffic on I-70 to dip from it. By 7:00 P.M. the state police had begun to complain to the Western Slope center about the danger.

By then, however, the air show was winding down. The last of the tankers departed by about 7:15 P.M. The helicopter pilots, complaining of smoke, ceased operations shortly after that.

The wind slackened, and the fire banked. The once-mighty smoke column began to dissipate. The lingering smoke made it impossible to tell how many acres had burned—estimates at the time ranged from one hundred to two thousand acres; one watcher at Canyon Creek Estates told a reporter that his guess was "a gazillion acres." Flames burned to within three hundred yards of homes, but no man-made structure was damaged and no resident injured.

NaVeaux, much bruised during his scouting flights, turned his plane for home. He instantly found himself in calm air; the plane's engine purred steadily for the first time in hours. NaVeaux had longed for such a moment, but when it arrived, it left him feeling empty. What he truly wanted, he realized, was to be back in the thick of things.

Ferneau, to the contrary, put Storm King behind him without regret. The air supervisor later told superiors he would happy to fly again for his home base in Idaho, but not for anyone in western Colorado, and especially not for the BLM's Grand Junction District. Until the day he retired, two years later in August 1996, he never did.

Petrilli and the other smoke jumpers emerged from their fire shelters at about 6:30 P.M., more than two hours after they had sheltered up. Their faces were coal-black.

Dale Longanecker came walking toward them through smoking trees. He had ridden out the fire sitting on a log near the place they had eaten lunch; by chance the fire had burned along the sides of the ridge but had spared the lunch spot, easily accessible to everyone who had been on the west-flank line. Longanecker, who mistrusted fire shelters, had left his behind early in the day after Sunny Archuleta, who had also left his, had assured him they would not need a shelter on a fire like this one.

Petrilli had reported by radio the names of the jumpers with him and heard back the names of the jumpers who could not be accounted for—Don Mackey, Jim Thrash and Roger Roth. He had tried without success to contact them. They could be safe, he told himself, in a spot where the terrain blocked radio transmissions.

"We're okay," Petrilli now told the pilots overhead, "but we're going to check on the other guys."

The jumpers began walking down Lunch Spot Ridge, intending to search along the west-flank line, but stopped when they saw pockets of unburned brush ahead. They had no desire to be caught by a fresh run of fire. They decided instead to conduct a grid search, fanning out on the flank of Hell's Gate Ridge and sweeping toward the peak of Storm King.

They confronted a scene of desolation. The once-impenetrable oak had been reduced to row upon row of spindly black trunks, punctuated here and there by skeletons of pine trees. The soil was dotted with shells of land snails, fallen in multitudes from the oak.

With most of the vegetation burned away, the jumpers had a clear view to the beginning of the west-flank line several hundred yards off. Even at that distance a series of horizontal bundles, small and low to the ground, drew their attention. The soft, crumpled shapes contrasted with the surrounding oak spikes; once noticed, they dominated the hillside. The jumpers abandoned the grid search and made a beeline for the place. After a few steps they came upon a piece of burned fire shelter.

They found the bodies of twelve firefighters clustered along the steep hundred-yard climb to the top of Hell's Gate Ridge. The bodies were so badly burned that dental records had to be consulted to make positive identifications. It took longer than that, much longer, to gather, sort and piece together the fragments of evidence that made up the last testament of the victims, the story of their final minutes.

For a while it looked as though, for decency's sake, the bodies would be taken away in haste and the record forever lost. But this was a fire of echoes, of actions where the past catches up to the present, for good or ill, and the memory of Mann Gulch was about to intervene again.

Beginning at approximately 6:45 P.M. Petrilli made a series of brief radio transmissions from the west-flank line. He reported finding a cluster of six firefighters near the bottom of the hundred-yard climb, then a space or gap as though someone were missing, then another cluster of five, and a few heartbreaking steps from the ridgetop, one last body.

Should a helicopter fly to the ridge for medical evacuation? Good asked from 93 Romeo, hovering overhead.

"No," Petrilli replied, "it's too late for that."

THE DISCOVERY OF the twelve bodies left only the two helicopter crewmen, Tyler and Browning, unaccounted for; the smoke jumpers who had arrived at the fire by bus began to organize a search for them. Survivors from the East Canyon had begun to show up at the meadow, raising hopes for Tyler and Browning.

Wayne Williams, a Missoula jumper, called the jumpers together in the meadow, out of earshot of reporters. Ken Wabaunsee, a member of the Flathead Native American tribe, was in charge, but Williams had more experience leading rescues.

"I told everybody to get on their knees like a football team so they couldn't see what we were doing," William said later. "Everybody thought we were praying, but it wasn't that."

The jumpers decided to fly by helicopter to the ridgetop and, if necessary, search through the night. Any who felt squeamish and wanted to stay at the meadow should raise their hands. Steele and Hurley looked at each other. "We take care of our own," Steele said, and they kept their hands down. A half dozen others raised theirs.

Archuleta, who had come to the meadow from Hell's Gate, began to organize the helicopter airlift. He took the first flight himself, aboard 93 Romeo, making a quick return to the place he had so recently fled. As he disembarked at H2, he found the survivors from Lunch Spot Ridge gathered there—Tony Petrilli, Dale Longanecker and the others, Mike Cooper, Mike Feliciano, Quentin Rhoades, Sonny Soto, Eric Shelton, Billy Thomas and Keith Woods.

"You guys can get a helicopter ride out if you like," Archuleta offered.

"No way," Petrilli said. Not in this wind, he told himself; not in a helicopter; not after what he had just lived through.

"Hey, that's fine," Archuleta said. "I just want to say you can get on this helicopter and get off this mountain if you like. If not, we're going to start ferrying jumpers."

The helicopter rocked in the wind a foot or two off the ground. As Petrilli and the other jumpers tried to decide whether to get aboard, a black snake five or six feet in length emerged from a hole and began to slither away. They let it go; anything that had survived the blowup, even a snake, had earned their respect.

When the rest of the search party arrived by helicopter, Archuleta pointed out where he had last seen Tyler and Browning as they apparently tried to circle above the East Canyon. Hurley led a group of searchers in that direction.

Williams and Steele separated from the others and walked to the top of the west-flank line. Their first impulse was to do whatever was necessary to remove the bodies from the mountain before nightfall as a token of respect. Supervisors at the meadow, with the same idea, radioed Good in 93 Romeo and told him to help with body retrieval.

It had been a long, long day for Good. Some of the bodies were Prineville Hot Shots he had flown to the ridge, "a beautiful bunch of kids"; he had seen nothing during a year and a half in Vietnam to prepare him for this.

"Hell no," Good replied.

Good ferried Petrilli's group of jumpers, with Petrilli reluctantly joining the last load, from the ridgetop to the meadow and put 93 Romeo on the ground for the night. The next day Good rented an automobile and drove to Denver. He eventually returned to fire work, but promised himself that in the future he would speak up more, offer an opinion about the fire, whether or not anyone wanted to hear from a helicopter pilot.

Williams and Steele walked the hundred yards of fire line where the bodies were located and then returned to the midpoint at The Tree. Two victims had deployed fire shelters near The Tree; the shelters were in disarray, partially stripped off. The tangle of bodies and shelters sent a clarion message: Something unusual and very wrong had happened here. Removing the bodies would be no act of tribute; it would destroy the opportunity to learn from what had happened.

"It dawned on me, we've got to leave this alone," Williams said later. "We had to leave everything picture perfect."

A natural curiosity, though, began to eat at Williams. Surely there would be no disrespect in trying to figure things out for himself, say, by raising the edge of a shelter to look underneath. Williams was about to make a move when a voice called out to him . . . from under the shelter.

"His damn radio went off, *calling my name!*" Williams said. "I jumped out of my skin. I was frantically reaching for my radio to respond to this call. By now people were starting to converge. It all hit me, and I started to throw up, but I didn't. I faked like I was coughing, and I don't think anybody noticed. Then I got my composure again."

The radio voice belonged to a supervisor in the meadow requesting an update. Williams reported in: The cold front had arrived, and they had no sleeping bags; Hurley and the others had ended their search without success; it was too dark for a helicopter to ferry them out. And, Williams told himself, he wasn't going to hang around dead men with live radios. They had no choice except to walk out.

A sliver of gold-and-crimson sunlight, seen through a haze of smoke, died out on the far horizon. Inky shadows filled the western drainage. In darkness the safest path would be the cleared west-flank line, and then the gutted bottom of the drainage. The jumpers began to file down the line.

Hurley was the last to leave the ridgetop. He switched on his headlamp. Its cone of light joined those from the lamps of other jumpers flashing across the terrible scene on the first hundred yards of fire line. Hurley waited a moment for his legs to stop shaking, and then he, too, went down the line.

CANYON CREEK ESTATES hosted a fire city by the time the smoke jumpers returned shortly after 11:00 P.M. Scores of firefighters had pulled in; engine crews had shown up from communities across the state—Rifle, New Castle, Basalt, Eagle and others. Residents had packed belongings into automobiles, but with the fire spreading away from the Estates, toward Glenwood Springs, most ignored the order to evacuate.

Artificial light spilled out of a garage that had been set up as a temporary command center for fire supervisors, a group becoming more numerous by the minute. Inside, men in unsoiled fire shirts and sparkling hard hats waved their arms and stabbed pointers at maps. The supervisors had plans for the returned jumpers. The twelve bodies had to come off the hill as soon as possible and without further discussion.

The jumpers made their way toward the light. Williams, Steele and Wabaunsee stepped forward into the unnatural glare. Their eyes were bloodshot. Sweat made black streaks in gray ash on their faces. Their clothing was rumpled and soiled. Behind them, the other jumpers muttered remarks, loud enough to be heard, about fire supervisors quick to issue orders but slow to dirty their hands.

"Those managers weren't the kind who'd spent a lot of time fighting forest fires," Williams said. "They were fat men in clean fire shirts who'd never thrown dirt."

Williams broke into the discussion.

"Listen, here's the deal," Williams said. "You move those god-damn bodies and you are going to ruin every bit of information those investigators can get. What happened up there was unusual, and it would be foolhardy to destroy that scene."

One of the supervisors said, archly, "Well, the governor wants those bodies off the hill tonight."

"Well, fuck the governor," Williams replied.

At this point a short, slightly overweight man stepped forward. His fire shirt looked as though it didn't belong on him; Williams thought he must be the garage owner. The man identified himself as Governor Romer and asked if there was a problem.

Williams, by all accounts, never missed a beat. He responded, according to a retelling by Romer himself, "Governor, be sure not to let them disturb that scene until there has been a very thorough mapping of what occurred, because in spite of the tragedy we must learn from this event. And there is much to be learned.

"Governor, please do what you can to be sure that scene is preserved, because I and others who in the future will risk our lives need to take advantage of the lessons of this day's event, tragic as it may be."

Something like this had happened once before, Williams said, in a place called Mann Gulch.

On that occasion the gathering of evidence had been done by untrained personnel. A crew of young smoke jumpers, told they would be helping with search and rescue, had wound up retrieving the bodies. They made no attempt to collect and document evidence, except to erect rock cairns where bodies were found.

"We weren't geared up for what we wound up doing," Homer "Skip" Stratton, head of the detail, said years later. "We broke over the ridge, and we could see the bodies in the burnt area. The snags were still smoking. Things were pretty rough."

As a result key questions would forever go unanswered: Exact,

undisputed locations had never been established for the fire that
Wag Dodge lighted and then stepped into, for the break in the
rimrock through which two of the survivors, Bob Sallee and Walt
Rumsey, escaped, or for large rocks where two fatally injured
men, Bill Hellman and Joe Sylvia, spent the night. Had Dodge's
fire blocked the escape for some of his crew? Probably not, but
that accusation lingered on with other doubts even now, forty-
five years later.

For Williams, Mann Gulch was more than a history lesson; it
was a living legacy. In the off-season Williams ran the visitor
center at the Aerial Fire Depot, the smoke-jumper base in Missoula that had been the starting point for the ill-fated drop into
Mann Gulch.

Three years earlier, in 1991, Williams had helped establish the
National Wildland Firefighter Memorial at the jumper base to
honor those who have died in wildland fires, especially those from
Mann Gulch. The memorial, an L-shaped stone wall surrounding
a pavement of granite bricks, recalls the Vietnam Memorial in
Washington, D.C.: The names of the thirteen Mann Gulch
smoke jumpers are carved into separate bricks in the pavement.
For the dedication ceremony, on May 8, 1991, Williams brought
together more than seventy family members, friends and others
with a tie to Mann Gulch, including Sallee, by then the last living
survivor.

When Williams finished, Romer replied that he was familiar
with the Mann Gulch fire from reading about it. Mann Gulch
had taught valuable lessons about human conduct and the forces
of nature, Romer agreed, and it would take a careful inquiry to
make clear the new lessons of Storm King.

Romer turned to the assembled fire managers. We'll do business the way Williams suggests, Romer said. Those who fell on
Storm King Mountain would be left, at least for a while, undis-
turbed.

Romer's handling of Colorado's disastrous 1994 fire season
won the approval of the state's voters. He was reelected that
November to a third term, becoming the Democratic party's senior governor; in 1997 he became a familiar face across the nation

as chairman of the Democratic National Committee. Romer recounted his experiences on the South Canyon fire at a memorial service for its victims held August 8, 1994, in Washington, D.C.; Williams was an honorary flag bearer at the service, in recognition of his actions.

THIRTEEN

DICK MANGAN ARRIVED home from work the evening of July 6 and, finding his wife and children out, proceeded to irrigate a pasture on his small, tidy, Black Angus cattle ranch located along the Clark Fork River a few miles north of Missoula. He worked during the day at old Fort Missoula, once an active military installation that now housed federal offices, including the Missoula Technology and Development Center, a division of the Forest Service, which since the 1960s has developed parachutes, protective clothing and other equipment for firefighters.

One of Mangan's duties as fire-program leader was to conduct investigations at the sites of fatal fires to check how the center's safety equipment operated under catastrophic conditions. He relaxed by working on his ranch, loving every minute, except when it came time to ship off the sleek black cattle, more pets than livestock.

When Mangan returned to the house at about 6:30 P.M., he found a message waiting on his answering machine. The dispatch office at the Missoula smoke-jumper base, the Aerial Fire Depot, wanted a return call right away. Such a message never brought good news, but this time Mangan felt better prepared than ever before.

That afternoon his equipment specialist, Ted Putnam, had announced the completion of a set of portable tool kits to gather and record evidence at a fatality site. The kits contained metal detectors, evidence bags, hundred-foot tape measures, magnifying glasses, tape recorders and notebooks—a regular Sherlock Holmes grab bag—as well as samples of materials burned at different temperatures for comparison purposes. Putnam had joked, "I've finally got our stuff organized—so we're almost guaranteed not to have an entrapment this year." Everyone had laughed.

Upon taking the job five years earlier, Mangan had determined that the center—known by its initials, MTDC—needed more in-depth information about the performance of its safety equipment. Although MTDC personnel had investigated fatality sites whenever possible, by 1994 they had never once arrived in time to observe bodies where they had fallen. The impulse to honor the dead by removing them from the scene had prevailed each time over the investigator's imperative, "Move nothing."

Mangan and Putnam had become more systematic in their inquiries over the years, adding the services of a Forest Service photographer, Jim Kautz, and more paraphernalia. The team had plenty of hands-on fire experience: Putnam and Kautz were ex–smoke jumpers; Mangan had been in fire work previous to joining the MTDC, and in 1994 continued to serve as an operations chief on national incident-command teams in times of crisis.

Mangan telephoned the Missoula dispatch office. "They didn't have any details, but something real bad had gone down. They said to start heading out."

As Mangan pulled his personal gear together, the dispatch office called back with an update. There were fatalities, perhaps more than a dozen, including a Missoula smoke jumper and several Prineville Hot Shots. Details remained sketchy.

Mangan hung up the phone and felt his world turn slowly upside down. In a few hours he would be on a blasted hillside with melted packs, charred Pulaski handles, crumpled fire shelters and, for the first time, victims. And for him they already had names and faces.

Prior to coming to Missoula, Mangan had been fire-staff officer on the Ochoco National Forest, headquartered in Prineville, which meant he had been administrative head of the Prineville Hot Shots. He had hired Tom Shepard and helped recruit Tami Bickett, a real catch, a fine college athlete who played basketball and volleyball. Supervisors were always on the lookout for women with superior upper-body strength; they called them the "boat people," because many came from rowing teams.

Memories flooded back. Other Prineville Hot Shots had gone to school with his children, Shannon and Patrick. Bonnie Holtby, a physically strong, hardworking girl, always trying to help someone else, had stayed with the Mangans while on a visit to Missoula with her sister, Stacy, one of his daughter Shannon's best friends. The Mangans and the Holtby girls had sat together at the kitchen table until the wee hours remembering Prineville days.

"I stood there and cried a bit," Mangan said, "and then the realization came home: We're the only ones to do this. There's nobody else who can go down there and do this investigation."

Mangan telephoned Putnam and Kautz and told them to meet him at the jumper base. They assembled quickly, but had to wait until 10:00 P.M. for a charter plane to fly in from Idaho. At first a few telephones jangled in the dispatch office, and then the ringing became nearly constant as sketchy news reports spread and firefighters, families and friends telephoned for details.

Mangan used the time to brief Putnam and Kautz: You should expect the biggest, most demanding event of your professional careers, he told them. "This one is going to be the Mann Gulch of our generation."

FIFTY MILES AWAY in the Bitterroot Valley, Nadine Mackey watched the late-evening news on the television set in her living room. The darkened picture window, which in daylight framed Blodgett Canyon, reflected flickering images from the screen. A fire in Colorado had blown up that afternoon, the news anchor said, causing fatalities. The number and identities of the dead and missing were not immediately available, but as many as fifty people had been on the mountain.

Bob Mackey came in from the kitchen.

"Bob," Nadine said, turning toward him, "isn't Don on a fire in Colorado?"

IN WASHINGTON, D.C., Forest Service Chief Jack Ward Thomas took a late-night call from a western field office. Be prepared to respond tomorrow morning to news that as many as forty or more firefighters have been killed on Storm King Mountain, the field office warned.

AN EERIE, END-OF-THE-WORLD scene greeted Mangan's MTDC team as their charter plane approached Walker Field in Grand Junction at about 1:30 A.M. A small brushfire was sweeping unopposed toward the airport runway. The commercial terminal was darkened for the night, but a pool of bright light glowed at a far end of the runway.

The MTDC team landed and found a telephone to report the brushfire. It took a while to obtain a rental van big enough for three men and a lot of equipment. By the time they loaded up, the brushfire appeared under control, or had burned itself out. They drove toward the light at the end of the runway, the Western Slope Coordination Center.

The center resembled a bunker under siege. The surrounding chain-link fence kept a fleet of news-media vehicles at bay. Beams of light poured from the center's windows. Inside, uniformed personnel with strained expressions dashed back and forth while telephones rang constantly.

Mangan, looking for a familiar face, hailed Mike Lowry, an old friend from Mangan's firefighting days in the Pacific Northwest. "It was total chaos," Mangan said. "Lowry and I had known each other for fifteen years, and he said, 'Come on, I'll brief you on what's going on.'"

Lowry already had done more than his share of briefing, gathering everyone at the center in late afternoon to pass on the first reports of fatalities. He had ordered crews hailing from the Pacific Northwest immediately relieved from fire duty in Colorado, concerned that they would become distraught and unmindful of their own safety once they learned what had happened. Then he had begun briefing officials outside the region.

Lowry had wound up using Chris Cuoco's cubbyhole to make the most sensitive telephone calls, the ones relaying names and numbers. Cuoco had first realized that something was wrong in early evening, when the BLM's Grand Junction dispatch office had telephoned a hurry-up request for a wind forecast for Storm King Mountain. Where the hell is Storm King Mountain? Cuoco asked himself, and dug out a map.

Cuoco returned the telephone call in minutes: Expect northwest winds with gusts of forty-five to fifty miles an hour, continuing through the evening.

"That's terrible, that's really bad," the dispatcher responded.

"What's this for?" Cuoco said.

"A helicopter rescue," she replied.

"I wish I had better news, but these winds are going to blast for at least a few more hours," Cuoco said.

Cuoco had been putting together his regular evening weather forecast when Lowry poked his head into the cubbyhole.

"Chris, I've got to use your office," Lowry said. "You do your forecast, but don't tell anyone what you're about to hear." Lowry, his voice uncharacteristically subdued, began reading a long list of names into the phone.

"I remember feeling like I was going to burst," Cuoco said. "I stared at the computer, but I couldn't move. Tears started rolling down my cheeks, and I pounded the desk. . . . I went back to writing my forecast, the headline of which was GUESS WHAT? NO RED FLAG WARNINGS!" Cuoco predicted cool temperatures and moderate winds, everything a firefighter could want except rain, for the next day, July 7.

Now, about 2:00 A.M. on the seventh, Lowry ushered Mangan, Putnam and Kautz into an empty office. He closed the door and by way of introduction said, "Things are really fucked up around here." That description applied to the whole operation, Lowry said, not just the immediate crisis. The dispatch system didn't work; the region had waited too long to call for help; people who should be playing on the same team openly feuded with each other.

Twelve bodies had been found, he said, and two additional firefighters were missing. Lowry was unable to say who the missing were at this hour, but the Garfield County sheriff was to begin a search at first light.

Lowry finished his account, and the MTDC team, who hadn't eaten for many hours, drove to a restaurant on the motel/fast-food strip near the airport. Breakfast seemed the most appropriate meal. When they finished eating, it was too late, or too early, for a motel bed, so they began the drive to Storm King.

The night sky darkened as they entered the narrows of De-beque Canyon, then softened as they emerged into the broad Colorado River Valley. Even in starlight Putnam knew where he was.

His last fire before taking the job at MTDC had been out there in the darkness on Battlement Mesa. He had been present on July 17, 1976, when a fire blew up in late afternoon and overran four firefighters, killing three of them, the last deaths from wildland fire in Colorado until Storm King. He had witnessed the death of one of the men, an event that had changed his life at the time and was about to change it again.

The Battlement Creek fire, as it was called, had made a nearly identical run the day before, July 16, in a nearby draw. By Put-nam's account the boss for that sector, Leonard Coleman, had ordered Putnam to take a young, inexperienced crew into a gulch Putnam feared was about to go up in flames. When Putnam refused, Coleman threatened to have him fired. Several of the youngsters protested that they weren't "chicken" and gladly would go down the gulch as ordered. Putnam screamed at them to start marching off the ridge. After they had walked a few hundred yards, the fire came roaring up the very same gulch they had been ordered to go down.

"Look at that," Putnam told the crew. "If you had your way, you'd all be dead."

The following day, July 17, the day of the fatalities, Putnam found himself on the same ridge, this time as boss of a sector next to Coleman's. The fire continued to churn below, but the overall plan called for everyone to quit work and find a safe zone by early afternoon. At about 2:30 P.M. Putnam was a mile from the fire eating a sandwich when he felt the air begin to vibrate. The fire was about to blow up again.

Putnam turned to a crew member, Randy Doman, and said, "Let's not take a chance." Putnam radioed Coleman in the adjoining sector and told him to be "heads up" because the fire was about to make another big run. Coleman radioed back that everything was fine, the crew in his sector had reported by radio that they were in a safe zone.

The smoke lifted for a moment and there, directly in the path of the onrushing fire, were four men. Putnam dropped his sandwich and grabbed his radio. At that moment a booming voice came over the radio—the fire boss, Jack Haslem, the man in overall charge.

"You dumb sons-a-bitches, get out of there!" Haslem shouted. "Get out, or I'm going to fire you."

"What the hell," Putnam told Doman, "if they don't listen to the fire boss when he's talking like that, they'll never listen to me." He put down his radio unused.

"What more could you do?" Doman said years later.

It turned out, however, that Haslem had been talking to someone else in a dangerous spot, and not to the four-man crew in the path of the fire, who heard nothing of the warning. The four included a squad boss, Tony Czak, and three other men, John Gibson, Scott Nelson and Stephen Furey.

The lone survivor, Gibson, speaking from a hospital bed a few days later, provided one of the few accounts in history of what it is like to be overrun by a blowup.

"All of a sudden the fire was coming at us. There's no more time for movement. Czak issued the order, 'Pull your vest off, water it down.' We each had gallon canteens. Water it down, water the back of your shirt and the back of your pants, get some water in the dirt, hit the dirt, holding your hard hat over your head.

"There's fire coming at us. You could feel the fire coming

over you and . . . it looked like we were doing all right. Then it got pretty hot, starting to hurt, and my pants started getting a little burned. So I ran my hand down my side and squirted some dirt on it.

"And all of a sudden Mr. Nelson yelled, 'I'm on fire.' He stood up and started to run. And Mr. Czak stood up and started to run. He was screaming. No words, just Ahhh, Ahhh, like this.

"I never looked up. I felt fire on the back of me, I scooped dirt up on me, and I was praying—Oh, I never prayed so fast in my whole life."

Gibson and Furey remained facedown in the fire line. Coleman rushed to the scene as soon as the blowup had passed.

"I saw Gibson sitting in the trail with his back to me without his hard hat," Coleman told investigators the day of the blowup. "I hollered to him, and he waved his hand in the air. When I reached him, he stood up.

"He had burns on his face, hands and arms but was coherent and able to talk—he pointed to Furey. I made Gibson sit down and looked at Furey—the back of his shirt and trousers was burned off. He was still breathing. I didn't touch him because of his burns.

"Furey stopped breathing, and I started mouth-to-mouth resuscitation. His lungs were full of fluid and smoke." Putnam and others had arrived by then. Furey was glassy-eyed. After a few minutes someone checked Furey's pulse and announced he had died.

Gibson, who suffered burns over 25 percent of his body, eventually recovered. When he learned of the events on Storm King Mountain eighteen years later, he asked, "Goddamn it, why does it happen again? After about one in the afternoon there ought not be people above these fires."

The official investigation of the Battlement Creek fire exonerated any individual from responsibility for the deaths, though it mentioned that Putnam, whose name was not used, had passed up an opportunity to radio a warning to the four-man crew. Instead the blame was placed on fuels, topography and unfortunate circumstances. Tragic misunderstandings put the four men in harm's way. Czak indeed had assured Coleman by radio that he and his crew were in a safe spot minutes before the fatalities occurred.

Coleman, contacted for this story, said he had tried to "forget the whole thing" in the intervening years and had no memory of threatening to have Putnam fired the day before the fatalities. "I honestly have no recollection of any discussion like that; I don't operate like that," Coleman said. A biologist by training, Coleman was assigned to the BLM's West Glenwood office when reached.

By Putnam's account Haslem, the fire boss, asked him after the fatal episode to remain silent about his earlier dispute with Coleman. That quarrel had nothing to do with the fatalities, Haslem argued; Coleman had suffered enough by having one of his men die in his arms. Haslem gave Putnam a "personal guarantee" that Coleman would never fight fire again.

Putnam accepted Haslem's promise and kept silent.

"At that time I didn't feel like 'Boy, all the truth of this has to come out,'" Putnam said. "It's only been in later years, when I've heard people talk about how the fuel combination and the weather helped to kill those people . . . Battlement Creek was very clearly the human factor thing, bad decisions. If you don't say everything, then you won't learn from what really happened."

Haslem, contacted in retirement, denied that he had asked Putnam not to repeat his story—he was contemptuous of the suggestion. When pressed on what he knew about the Putnam-Coleman quarrel, Haslem responded, "How the hell can I be expected to remember" events from that long ago.

The denials and memory gaps offer no support for Putnam's version of events, but his story should not be entirely discounted. He does not make himself a hero with the telling—he admits that he failed to warn Czak and the others, and that he failed to tell investigators what happened the day before the fatalities.

A final piece of this puzzle would fall into place in the weeks after the South Canyon fire, challenging Putnam's commitment to his story. Once again, as on Battlement Mesa, he would face a choice between insisting on his version of events or accepting a picture he knew to be flawed. And this time around, he would be one of the investigators.

FOURTEEN

THE NIGHT GREW cold. Orange flames blossomed and then died out on the slopes of Storm King Mountain above the sprawling fire camp at Canyon Creek Estates. Wisps of smoke drifted through the camp. A Glenwood Springs ambulance pulled away, siren blaring for no apparent reason except a thwarted sense of emergency; a truck marked IDAHO DEPARTMENT OF LANDS pulled in. Television-news vans sprouting satellite dishes carried fantastically multiplied images of the scene to an outside world.

Exhausted firefighters from reinforcing units tried to bed down. They moved as shadows, their uniforms, rank and gender obscured by darkness. The mountain became its own night sky, a dark mass glowing with constellations of tiny fires connected by swirling gases in imitation of stellar matter. The real sky formed a black roof, then brightened as the Milky Way emerged. A television newsman, struggling with his report, told viewers that the news would be tragic if the scene "weren't so beautiful," as though the beauty of the night had prevailed over the events of the day.

By 2:30 A.M. on the morning of July 7, Robert Nash, a food-van driver, had delivered 260 meals to the fire camp. Food checkers and volunteers at the Safeway and City Market supermarkets in Glenwood Springs stuffed sandwiches and cans of juice into bags; Nash was their window on disaster. "How's it going?" they asked when he came by, but he had nothing new to report. "I'm concerned because the men have got to drop from exhaustion," said Mary Casteel at the City Market.

It was late at night before Brad Haugh left the BLM's West Glenwood office, where he had remained when fire appeared on nearby hills. Haugh made his way to the Valley View Hospital for treatment of a slight burn on his left elbow, received in his dash over Hell's Gate Ridge ahead of Kevin Erickson. He had waited this long because he thought the hospital emergency room would be crowded with more seriously injured firefighters. The place was empty when he arrived; that was bad news, he told himself, very bad.

Once bandaged, he mounted his motorcycle and cruised the main drag of Glenwood Springs, gunning the engine, running stop signs and blowing red lights, the throb of the engine setting his body tingling. Goddamn, I'm alive! he told himself. I'm alive!

Sleep did not come easily to the two survivors admitted to Valley View Hospital, Eric Hipke and Kim Valentine; Erickson had been released after treatment for second-degree burns on both elbows. Hipke's burns covered 10 percent of his body and were worst on his hands and arms. Bandages made it impossible for him to turn over. An intravenous morphine drip gave him flashy nightmares, interrupted from time to time by Valentine's sobs from a nearby room.

Bill Baker, an emergency medical technician as well as a Prineville Hot Shot, had accompanied Valentine to the hospital and stayed on afterward. Baker was in the reception area when Butch Blanco came by to check on the injured. Blanco, appearing dazed, recognized Baker as a firefighter from the mountain.

"First thing Blanco said, he goes, 'What the hell happened? What in the hell happened?'" Baker, taken aback, replied with a question of his own.

"You're Blanco, aren't you? You're the incident commander, right?" Blanco should have the answers if anyone did.

"What the hell happened?" Blanco repeated.

"The damned thing just blew up," Baker said.

Later a nurse requested that Baker visit Hipke, who was beginning to ask questions of his own; Baker went to Hipke's room. "All those guys made it, didn't they?" asked Hipke, who still thought that the others from the west-flank line had crossed the ridgetop somewhere behind him. No, Baker replied; half the Prineville Hot Shots and several others were missing and presumed lost. Hipke, woozy from drugs, began to choke up. An embarrassed Baker withdrew from the room.

Hours later, at half-dawn, Hipke was awakened by a shaking of his hospital bed. The room furniture began to rattle. Oh, no, Hipke thought as the hospital's laundry ground to life on a floor below, I was just in a fire . . . and now I'm in an earthquake!

THE MTDC INVESTIGATORS drove past Storm King Mountain before first light, at about 4:30 A.M., and located the command post at the Glenwood Middle School, where it had been shifted to from the garage at Canyon Creek Estates. After a briefing there they proceeded to the fire camp at Canyon Creek Estates to await a helicopter ride to Hell's Gate Ridge.

"It was a bizarre experience," Mangan said later. "Two lanes of the interstate were closed. There were flames on the hills outside town; it was smoky; there was media everywhere. They were bringing in a national management team, trying to make things happen, because they thought the town was threatened. And it was."

The fire camp was coming to life as they arrived. The doors of ambulances, fire engines, panel trucks and pickups swung open, and men and women stumbled out. Figures in yellow shirts and green pants struggled off the ground. They mingled together holding coffee cups and began to gather tools and gear.

At 6:52 A.M., by the watch of Nash, the food-van driver, the *buzz-whop* of a helicopter announced the commencement of the day's air show. An enormous chopper hove into view with a fifteen-hundred-gallon water bucket suspended from its belly. Two Delta County School District buses pulled in, loaded with firefighters. A truck from Roaring Fork Valley Co-Op rumbled in with a load of retardant liquid. A half dozen ambulances, a reminder of the grimmest business of the day, lined up alongside a roadway.

Thursday, July 7, the sixth day of the South Canyon fire, should have been a mopping-up operation, a time to clear away the aftermath of disaster, restore a sense of control and open an investigation. The chronology of events in the South Canyon Fire Investigation has no entry for July 7 or for any later day. In the official record nothing occurred after the day of the blowup, July 6, to merit making into a lesson for the future. On July 7, however, serious challenges remained: Two firefighters were missing, the fire burned on, and by the end of the day a public outcry would arise over what seemed more and more to have been a preventable loss of life.

The rising sun disclosed a mountain covered with leaden ash and smoldering tree trunks. The fire had spread to more than two thousand acres, more than three square miles; it sputtered to life under a warming sun, flaring in clumps and patches, once again a threat to homes, property, lives—and reputations. In Grand Junction the Western Slope Coordination Center logged a directive at 8:40 A.M.: "South Canyon fire gets all resources they want."

Before the day was over, more than 240 firefighters would battle the flames. An air show including six tankers would "paint the hillside pink," in the words of an air supervisor; helicopters would dump hundreds of thousands of gallons of water and haul hundreds of people and tons of gear. A fleet of at least seventeen fire engines would protect lives and property. Heavy earthmoving equipment—including five bulldozers, one of which had to be flown in—would crawl the slopes. Even nature would help: The weather turned cool and windless, as Cuoco had forecast. But when July 7 ended, the fire burned on, and Tyler and Browning remained missing.

A plan had been drawn up overnight to deal with the "incident within an incident," as the search for Tyler and Browning came to be known. Steve Crockett, a Colorado emergency-services official and former Aspen fire chief, had been called to the scene to be head of planning—people from a variety of local agencies who had fire experience had been hastily assembled after the blowup to run the fire. At 2:34 A.M. on the seventh, Crockett produced the "incident objectives" for the day, listed in order of priority.

The first priority was the safety of those conducting the search-and-rescue operation. Number two was to "confirm location and status of all firefighting personnel," a way of saying, according to Crockett, Go find the people, dead or alive. The protection of homes and other property ranked number three, and only then, in the fourth position, came fighting the fire. Crockett's plan matched standard guidelines, which rank protection of life as first priority, protection of property next and forest values last.

By law, responsibility for the search-and-rescue mission fell to the county sheriff. Never mind that the search was for federal employees, that the BLM and Forest Service had more resources on hand than the sheriff and that the sheriff had other, pressing duties to perform—namely, evacuation of homes, security in evacuated areas, traffic control and body removal. Levy Burris, the Garfield County undersheriff, took charge of the search mission using a volunteer group, Garfield County Search and Rescue.

Trouble started almost immediately when the BLM, according to Crockett, refused to allow its helicopters or personnel to be used in the search. A decision had been made at the BLM's Grand Junction District, Crockett was told, to commit federal resources to protect homes and put out the fire, not to conduct a search. "Sometime in the night, about three A.M., Winslow Robertson said, 'We can't support your plan,' " Crockett recounted.

Crockett asked Robertson if he could use 93 Romeo, parked in the meadow. Robertson replied, "That's gone, too . . . don't go through us, go through your own channels." Robertson appeared to be passing along decisions made by others, Crockett said.

Robertson later told fire investigators that the search-and-rescue operation had been "botched, handled badly," but he did not elaborate. Pete Blume did not address the issue in interviews with investigators. Blume and Robertson both declined repeated requests for personal interviews for this book.

Burris, upon learning that the sheriff's department was on its own, ordered a helicopter from a private company in Glenwood Springs. When that chopper landed at the meadow in the morning, however, it was found to lack proper safety clearances. "This was not the time to throw out the rule book; it was time to play by the rules," said Crockett, who refused to allow it to go into action.

The leasing company sent a replacement helicopter, but by then fresh objections had cropped up to having any private craft use the meadow landing zone, a federal facility. Pat Basch, Tyler's old co-foreman and rival, had been placed in charge of the meadow helibase and was trying to enforce federal standards for machines and personnel. Basch first had insisted that the search-and-rescue volunteers be equipped with boots, fire clothing, hard hats and earplugs.

Then, when the second leased helicopter showed up, it was discovered that the *pilot* lacked proper clearances. Basch refused to allow it to land.

Levy Burris, the undersheriff, became enraged.

"We had pissed away three hours or better, and then he [Basch] tried to tell me we weren't going to land there," Burris said later. "And I said, 'I'll have a deputy put you in handcuffs to get you out of my way.' And I had a deputy in that helicopter, too."

Everyone realized that the situation was getting out of hand. Rule books were abandoned, and the helicopter came to earth at the meadow helibase.

In the meantime, Dick Mangan and Ted Putnam, waiting to be ferried to the ridgetop, had become much alarmed by the failure to begin the search-and-rescue operation. The chances of survival for a victim of serious burns without medical attention decline dramatically over the first twenty-four hours. The window of survivability was shutting down for Tyler and Browning.

The BLM helicopter, 93 Romeo, continued to sit in the meadow, awaiting a required servicing. It was 11:00 A.M. before it was flown to Grand Junction, serviced and returned.

One of its first tasks was to shuttle the MTDC team, after more than a five-hour wait, to the ridgetop. The team disembarked and broke out the new investigation kits. Putnam began to concentrate his thoughts, trying to shut out everything except the job at hand.

Putnam had found that meditation, focusing on the immediate task, helped him handle the stress of a fatality site. An interest in experimental psychology, a field in which he held a doctorate from the University of Montana, had led him several years earlier to study Buddhist psychology and its meditation techniques, and ultimately to take Buddhist vows.

"When I was on the Battlement Creek fire, I thought of the actual people," he said. "It doesn't take long before you're a nervous wreck. When I go to a fatality site now, I've had some meditation training, 'mindfulness' training, keeping your mind on what you're doing and not letting it wander. As the first shock set in, I told myself to go back into that mindfulness thing."

This time meditation failed him. As Putnam started down the fire line, he came upon the first body, the one separate from the others near the ridgetop. None of the bodies had been identified, so Putnam gave them numbers, Firefighters numbers one through twelve.

Firefighter Number 1 was a large, muscular male located forty yards below the ridgetop. He was kneeling in the formal posture, composed as a statue, of a Muslim at prayer. His head was bowed to the ground, and his hands were lying flat on either side of his head. He faced east, in the direction of a personal, unobtainable Mecca, the top of the ridge.

The posture implied that Firefighter Number 1 had been overtaken by an unexpected wave of flame and then walked on a few steps, gradually "winding down." The burn marks on his clothing were consistent with that explanation, except for one remarkable detail—the boot laces were missing.

It is not unusual for a fire to spare footgear. A racing fire leaves a layer of relatively cooler air a foot or more above the ground, at boot height; the phenomenon is common enough that firefighters are trained to take advantage of it in extreme situations by dropping to the ground.

Nor is it unusual for laces and boots both to be burned; if a flame front comes after the first wave of fire, it burns to the ground. But never before had Putnam seen boot laces burned away while boots remained intact.

Firefighter Number 1's clothing showed a pattern of higher temperature farther up the body, as would be expected if he had been standing upright when struck by the fire. A few bits of yellow shirt remained intact, indicating temperatures of 800 to 900 degrees. An intact nylon belt indicated a temperature below 470 degrees. A strip of blue T-shirt around the waist, perhaps protected by the belt and pants, indicated a temperature below 430 degrees.

Searing heat must have reached ground level, but only for a moment or two, to burn the laces. Two canteens found near the body told the story: One, filled with water, was intact, while the other, empty and thus easy to heat, was partially melted. A few tufts of grass and leaves remained unburned. Scattered food cans had not popped open. It was as though the fire had slapped the ground and rolled onward.

Another ten feet downhill Putnam found a fire shelter inside its plastic case. Why had no attempt been made to deploy it? A trained person can shake one out and get inside in twenty seconds. In Putnam's opinion Firefighter Number 1 was high enough up the slope to have survived in a fire shelter. Two people had temporarily survived roughly equivalent temperatures in Mann Gulch in the days before fire shelters. Bill Hellman, the squad boss, and Joe Sylvia had lived through the night in the gulch after being overrun by the blowup; they died in a hospital in Helena the next day, less than twenty-four hours after being burned.

A thought broke Putnam's concentration: What if the two missing men on Storm King had been injured as Hellman and Sylvia had been and were desperate for help? It was now around noon, twenty hours since the blowup; the twenty-four-hour window of survivability remained open, though barely.

Putnam made an initial survey of the eleven other bodies: Nothing added up. This stretch of the fire line commanded a good view of the western drainage. The victims should have seen the fire well before it arrived and prepared by deploying fire shelters. But Putnam found only two fully opened shelters. Those deployments had failed to save anyone, but why only two out of twelve? Why hadn't there been more?

Most baffling of all, the bodies lay in an orderly line, as though shot from behind. Except for Firefighter Number 1, none appeared to have moved after being struck by the fire. On other fatal fires the victims made a dash or scramble after being struck by flames, unless they had deliberately dropped to the ground beforehand. They would wind up in random, scattered spots, which is what happened on the Battlement Creek fire. Two of those victims, Tony Czak and Scott Nelson, were found dozens of yards from the others.

The fire on Storm King had struck with a ferocity beyond Putnam's experience. Someone asked him at this point what he was thinking. "I'm not sure what happened, but it's different from anything I've seen in the past twenty years," he replied.

With nothing about the dead making sense, Putnam's thoughts turned again to the living, the missing firefighters. "Where the hell is search and rescue?" Putnam said, first to himself and then out loud.

He knew he had no business leaving the fatality scene; he alone might recognize the importance of evidence uncovered when the twelve bodies were moved. But there would be time later to gather artifacts and analyze what had happened. The photographer, Jim Kautz, could record the body-removal operation.

Putnam broke off a tape recording of his observations with the comment that as of this moment the living mattered more than the dead. He went to find Mangan and commence a search.

FIFTEEN

THE AIR WAS cool, but the slopes of Storm King were hot and dry under a midafternoon sun. "A search wasn't our primary job, it wasn't our mission," Dick Mangan said later. But if the missing firefighters were alive, waiting for help, they likely had no water. A picture of Bill Hellman in Mann Gulch kept flashing in Mangan's mind: During the night, the injured Hellman had drunk from a can of Irish potatoes, the briny liquid giving him a raging thirst. "I think all of us can imagine in our worst dreams what that would be like," Mangan said.

He and Putnam began searching in the most obvious place, the East Canyon, but found that they had stepped into a blind alley. Deep ravines appeared unexpectedly and snaked off in twists and turns. The searchers lost sight of each other and the horizon. It wound up taking forever to check the smallest area, but a telltale smell held them in the canyon. "Something had died in there," Mangan said. It was nearly 5:00 P.M. before they located the source of the smell—the carcass of the long-dead deer or elk.

About a dozen sheriff's volunteers had arrived on Hell's Gate Ridge in the meantime, but with no information about where the missing men might be. Survivors who could tell them had been gathered away from prying eyes and microphones at the Hilton Hotel in Grand Junction.

Half the sheriff's volunteers began to search near the west-flank line, while the remainder began the sad, necessary task of carrying bodies to the ridgetop, where a helicopter could pick them up. Every hand was needed. The United States congressman for the district, Scott McInnis, and his wife, Lori, had asked to be flown to the ridge, and both immediately pitched in. So did an observer for the Federal Emergency Management Agency, Andrea Booher.

"It was surreal," Booher told a *Glenwood Post* reporter. "Everything was chopped low to the ground, and whatever was not chopped was charred. The earth was very, very hot; you could feel it through your feet. There was one man near the top, and every time I went by him, I thought, 'If only he could have made it to the top.' "

This burst of activity appears to have generated a false report that the body of a thirteenth firefighter had been found. At the time the body-removal operation began, about 3:00 P.M., Patty Tyler was awaiting word of Rich at home in Palisade, near Grand Junction, with her son, Andrew, whose first birthday was coming up on July 12. She had learned in the morning through a network of friends that the BLM privately listed Tyler as a "presumed fatality." Patty, a public-health nurse, thought that judgment irresponsible without the evidence of a body; she continued to hold out hope.

At about 3:30 P.M. Mike Mottice, the resource manager for the BLM's West Glenwood office, telephoned Patty and told her that Rich was a "confirmed fatality." Mottice's report sounded final, so two hours later Patty was shocked to hear a television-news report that Tyler and Browning were still missing and unaccounted for, not dead. She made a note of the time, 5:51 P.M., and telephoned Mottice to demand an explanation.

Mottice apologized, said the information had been passed on to him by others and blamed the BLM's lack of experience in notifying next of kin. "I'm really sorry, but we've never done anything like this before," Mottice said.

"Well, neither have I," Patty replied with fury. "These are

families! These are real people! We need to know what's going on."

By 5:00 P.M. Mangan had left Hell's Gate Ridge and taken a helicopter shuttle back to Canyon Creek Estates. There he made contact with the national incident-management team, twenty-seven seasoned fire managers flown in by NIFC to take over supervision of the South Canyon fire. The team had been pulled together in California, wildfire's glamour state because of its many spectacular brushfires in populated areas.

Mangan located the team's IC, Jack Lee, who turned out to be a former supervisor of Mangan's, and asked for a helicopter. "We've got to have it right now. We've got to go find those people," Mangan said.

Lee and his team had arrived at Grand Junction during the night, about two hours behind Mangan's, and had met with Pete Blume at the BLM district office for a briefing. "It was emotional; the two missing helicopter crewmen were Blume's men," Lee said later. "My direction was to get to Glenwood Springs immediately and stabilize the situation."

Blume told Lee that the national team would be in charge of fire suppression on Storm King, while the Garfield County sheriff would be responsible for the search-and-rescue mission. Lee should provide assistance to the sheriff if he asked for it. Lee, who kept a daily journal of his activities, noted that Blume told him, "No. 1 priority was to support Garfield Co. SO [sheriff's office] in body recovery and [Mangan's] investigation team. We have fire control, they the burnover incident."

The arrangement covered the bases, but it called on Lee to play second fiddle to a local sheriff while running a major fire, ensuring minimal attention to the sheriff's problems. This happened about the time the BLM told Steve Crockett, the plans officer at Storm King, it would not help with the search-and-rescue mission.

There is a likely explanation for the BLM's attitude. The agency appears to have written off Tyler and Browning's chances of being found alive as early as the night of the sixth, and certainly by the next day, when Tyler was declared a presumed fatality before his body was found. Lee's notes of his meeting with Blume the night of the sixth do not mention the word "search," but instead refer to "body recovery." There was in fact a body-recovery operation under way that night, until Wayne Williams and Governor Romer stopped it, but no search by the BLM, not that night and not the next day.

What could be clearer? Blume was telling Lee to help with body recovery, not with a rescue mission regarded as a lost cause.

Lee quickly became occupied with the problems of taking over and running a two-thousand-acre fire. Blume had told Lee to relieve "as soon as we get there" the local supervisory team pulled together after the blowup. The quick takeover by the national team was a switch from the normal procedure of a gradual transition during a twelve-hour shift. The BLM leadership in Grand Junction, Lee later said, had doubts about the abilities of the local team it had created.

The transition turned into a nightmare. Crews showed up late and without tools; so many VIPs came to the scene that Lee made a list to keep their names straight. On top of that, the "temporary" local team was in no hurry to relinquish authority.

"Nobody had told them the transition would be quick," Lee said. The local team thought that the national team was trying to push them around and, worse, had abandoned the search-and-rescue mission, the local team's top priority.

Steve Crockett, the local plans officer, felt insulted and left out. "Their attitude was 'We're from California, we're here to save you. You screwed it up.'"

At one point a member of the national team tapped Levy Burris, who as undersheriff was running the search-and-rescue operation, with the antenna of his two-way radio to inquire who he was. The plans officer, Crockett, recounted, "I'm standing there with Levy, and this guy from the incoming team starts poking Levy in the chest. I'm going, Well, *this* ought to be interesting. Levy's a short, stocky guy; he doesn't have a reverse in his gearbox.

"The guy says, 'I'm up here to find out where the media's getting all this information about bodies, where the leak is. I want to know who is in charge here.' Levy says, 'That's me; I'm the undersheriff.' Oops. You could hear the federal guy knock out gear teeth trying to get it into reverse."

By the day's end Crockett asked to talk privately with Lee to air his complaints. "I was at the drool stage," Crockett said. "All they could think about was body recovery, nobody was searching. I couldn't get anybody to listen to me: It was like those ads for the Philadelphia *Inquirer* where the guy's going around and can't get people to notice a safe is about to fall on them."

Crockett wound up screaming obscenities at Lee. By Crockett's account, he said, "I hope you sons-a-bitches never have to come back to this part of the country again." By Lee's account, Crockett's language was more vulgar and personally insulting.

"I should have hit him," said Lee, who noted in his journal, "Local hard ass named Steve Crockett . . . Very unprofessional and loudmouth. Ranted and raved, made no real sense. Will talk to Levy." Crockett was told not to return to fire camp.

Lee had come to his own conclusion that Tyler and Browning would not be found alive. "I looked at that mountain, and it was a moonscape, it was nonsurvivable," Lee said later. "The mission should have been called 'search and recovery,' not 'search and rescue.' "

But Lee did not refuse assistance to searchers. When Mangan requested a helicopter on the afternoon of the seventh, Lee found him a National Guard Huey. Mangan flew back to the ridgetop at about 5:30 P.M. and picked up Putnam. They confined their aerial reconaissance to the East Canyon and Lunch Spot Ridge, certain that the missing men would be found there.

Fire had blackened the slopes; every fallen tree could be a fallen firefighter. Mangan and Putnam ordered the pilot to make passes closer and closer to the ground. In Mann Gulch the body retrieval detail had been unable to locate the last victim until they realized they had been walking past the body all day, mistaking it for a tree stump.

As the National Guard Huey dipped and rose above Storm King, thermal drafts from the heat-soaked ground buffeted the craft. "If the engine had coughed once, we would have crashed," Putnam said. At one point the searchers spotted what looked like a body, but upon landing discovered a charred log with branches sticking up.

After two hours, at 7:30 P.M., with plenty of daylight left, Putnam was ready to quit. Those conducting rescue missions often face a difficult choice between concern for victims and concern for themselves, and veterans know that the heroic role, risking everything for others, is not always the right choice. Putnam slapped Mangan on the shoulder and made a throat-slashing gesture. They ordered the pilot to head for the helibase in the meadow at Canyon Creek Estates.

By the time they landed, the South Canyon fire had risen to the status of a national disaster, attracting political figures from as far away as the nation's capital. Bruce Babbitt, secretary of the interior, had flown in during the night; he kept a low profile, declining to speculate about Storm King except to say that fire-fighting was a dangerous business. Once a hotshot himself, Babbitt had been in Colorado a few days earlier, picking up a Pulaski and joining a fire crew in a show of support.

Jack Ward Thomas, chief of the Forest Service, arrived from Washington and joined survivors at the Hilton Hotel wearing an old pair of White Boots and work clothes. He was the first biologist ever to head the agency, a sign of changing times, but had shoveled dirt on many fire lines. He stood several rounds at the bar. "They're hurting; they need to go home," Thomas said. "They're gutsy and tough; they're tight-knit people."

President Clinton telephoned Governor Romer from Air Force One while on a European visit to offer encouragement and federal assistance. Romer quoted Clinton as calling the firefighters "gutsy and courageous," and saying, "You certainly have it in hand as best you can, but please let us know anything else we can do."

THURSDAY, JULY 7, had begun with clear objectives—find the missing men and control the fire—but headed toward an exhausted conclusion with neither goal accomplished and amid a growing public chorus of recrimination. Everyone from homeowners to firefighters asked, Who is to blame? Many thought they knew.

Jan Girodo, whose family had been forced to evacuate their home in the Storm King Trailer Court, expressed a common bitterness: "They said they were keeping an eye on it; then people were dead. That's what we don't understand. Before this is over, people are going to be calling it the Bureau of Land Murderers."

Even those who arrived after the blowup voiced criticism. "If you're a hotshot and you die, then there's something wrong, because they are experienced firefighters, they know what they're doing," said Matthew Black Eagle, a Sioux from Montana on a forty-person crew assigned to Storm King on the seventh. "This is the biggest cluster-fuck I've ever been involved with."

News reporters asked the assembled political figures for answers and met a surprisingly common front. "I don't think now's the time to start pointing fingers of blame," said Governor Romer, who began referring to arguments he had heard from Wayne Williams the night before, when Williams convinced him to leave the bodies where they were. "Before you move any bodies or any equipment, you've got to thoroughly investigate this," Romer said, echoing the Missoula smoke jumper. "We must learn from this incident. . . . It can save lives in the future."

Representative McInnis came down from the ridgetop with a dramatic description of the fatal site and a message similar to Romer's: Hold off judgment until an investigation is complete.

"It was like opening the gates of hell," said McInnis, a Glenwood Springs policeman before running for Congress. "Anybody that is second-guessing what went on up there doesn't know what they're talking about. Those young men and women are heroes. They had no chance, no chance. What happened was an accumulation of unforgiving circumstances. That's what got them."

Taken together, the calls for self-restraint by Romer, McInnis and others cooled the public debate, and that was no small accomplishment.

The need for an investigation had been well anticipated; the Forest Service and BLM had agreed to a cooperative effort within a few hours of the blowup, and on the seventh the heads of both agencies—Thomas for the Forest Service and Mike Dombeck, acting director of the BLM—signed an agreement to pursue a joint inquiry. A team began to assemble in Colorado for that purpose. The members were nearly all federal employees actively involved in fire work, which limited their outlook and the amount of time they could devote to the investigation, but the leader, Les Rosenkrance, had authority to appoint additional members. The team was to submit a report with recommended actions within forty-five days, a deadline agreed upon by compromise.

As night fell, Lee and his incident-management team began the more immediate task of setting priorities for the next day, Friday, July 8. Lee's incident objectives, completed at 1:00 A.M. on the eighth, omit any mention of the continuing search for the two missing helicopter crewmen, with the notable exception that priority number one calls for "rescuer safety," an echo of Steve Crockett's list from the previous day. But no one is assigned rescue duties; the other priorities concern protecting property and fighting the fire.

Mangan and Putnam say that Lee promised to provide them with Type I crews and another helicopter to resume the search the morning of the eighth, but no such resource order was made. Lee has no memory of making the promise. Lee was a busy man, and had provided a helicopter the previous afternoon, so it is fair to assume he simply forgot his promise.

It is also fair to say that the failure of the search-and-rescue effort on the seventh should be a lesson, and an embarrassing one, for the federal government and for the BLM in particular. Tyler and Browning might have survived for hours, though their chances were slim. A greater effort could and should have been made to find them, especially during the first critical twenty-four hours. The fire, after all, did not occur in a remote wilderness, as Mann Gulch did; it burned next to an interstate highway.

The possible survival of Tyler and Browning was the main issue, but not the only one involved. The failure to find the two men left their families in doubt about their fate far longer than necessary. Patty Tyler in particular was sent on an emotional roller-coaster ride that amounted to torture.

When the BLM withheld support for the search, the burden shifted to others not as well equipped as the BLM. The Garfield County sheriff's office, despite drained resources, made an effort with volunteers and a leased helicopter. Mangan and Putnam abandoned their duties, perhaps wrongly, to conduct their own search. The BLM by contrast did not even try.

The lesson: If a search is required for federal employees on federal lands, there is a moral if not legal responsibility for the federal government to provide more help than was done on this occasion, especially if federal workers and equipment are standing by at the scene.

THE HILTON HOTEL in Grand Junction, located on motel row just off I-70 near Walker Airport, is an oasis of shaded patios, green lawns and a swimming pool. Over one hundred firefighters gathered there July 7 as crews from around the region joined the survivors of the South Canyon fire. The Zig Zag Hot Shots, who had camped with the Prineville Hot Shots in Fruita the night before the blowup, had been fighting a fire elsewhere in the region when they heard the news. They put down their tools and refused to work until the IC agreed to send them to Grand Junction.

There, the quest for accountability had begun, extending even to the weatherman, Chris Cuoco. Shortly after the first news of fatalities, the National Weather Service had ordered Cuoco to supply copies of his forecasts for review by the national director's office in Washington. Cuoco's predictions stood up to scrutiny. Over breakfast, however, Cuoco had learned for the first time, by reading a newspaper account, that his Red Flag Warnings had never reached Storm King. He wept.

The first newspaper stories were sketchy, and as the day wore on, pressure mounted for more information. Jack Lee, reacting to what he termed a "media frenzy," held a press conference at 1:00 P.M., though he had little new to add. "Got my ass beat up for not knowing all the unknown details," Lee noted in his journal. "Did not have names, a description of what had happened or who was not found."

Embarrassed and angry, Lee ordered his staff to obtain immediate written confirmation that families of the fourteen missing had been notified so that the names could be publicly released. Teams of federal officials had been sent to the homes of many of the twelve known dead beginning the night before.

IT HAD BEEN nearly midnight when Walt Smith, a Forest Service supervisor who had grown up with Don Mackey, received a telephone call from regional headquarters in Missoula asking him to drive to the Mackey home in the Bitterroot Valley. Smith, who had helped train Mackey as a smoke jumper and had jumped many fires with him, was a bluff, stern man, but when he was around Mackey, he changed. Mackey had been able to make him laugh; Nadine Mackey remembers overhearing them together, giggling.

Smith already had been on the telephone with Bob and Nadine, sharing what little he knew. Smith and another Forest Service supervisor, Joe Wagenfehr, pulled up at the Mackey house at about 1:00 A.M. on the seventh. The Mackeys knew that something was wrong when they saw Wagenfehr, a stranger to them, get out of the car. Everyone went into the house.

"I'm sorry, Don died," Smith said. Nadine collapsed as he was speaking.

WITH JACK LEE'S prodding, the list of fourteen names was read to the press in Colorado at 4:15 P.M. on July 7.

Prineville Hot Shots

KATHI BECK
TAMI BICKETT
SCOTT BLECHA
LEVI BRINKLEY
DOUG DUNBAR
TERRI HAGEN
BONNIE HOLTBY
ROB JOHNSON
JON KELSO

Smoke Jumpers

DON MACKEY
ROGER ROTH
JAMES THRASH

Helitacks

RICHARD TYLER
ROB BROWNING

By then fire managers had decided to hold a "critical-incident debriefing" at the Hilton for everyone who had worked on the fire. The remaining Prineville Hot Shots, wearing blue T-shirts with the dancing-coyote logo, gathered in the hotel conference room expecting, as did others, the first detailed account of what had happened. A crowd filled the seats and lined the walls. "There were only fifty people on the fire, but there were nearly a hundred and fifty people in that room," one participant said.

When the meeting came to order, it turned out that the "critical-incident debriefing" was not about the events of the fire. Instead a clinical psychologist, instantly nicknamed "Sigmund," was to conduct a grief-counseling session. Sigmund began to lecture about the grieving process, flipping through charts to explain the five stages of the "grief cycle"—denial, rage, bargaining, depression and resolution.

The firefighters shot glances at each other and began muttering, "This is bullshit." The Prineville Hot Shots asked questions about the fire, but Sigmund had no answers. Finally a supervisor with a sense of mercy interrupted and announced that further attendance was optional. The Prineville Hot Shots led a walkout.

As the crowd spilled out of the conference room, the survivors drew together on the patio, in hallways and in the bar. Smoke jumpers recounted what it had been like on the west-flank line. Others compared memories, trying to piece together who had been standing where, who had said what, taking the first steps on a long journey into the past.

Sigmund was fired on the spot by Mike Lowry, who had hired him. "That was the poorest job of anything I'd ever seen," said

Lowry. "I told him that in so many words. The fire community is very small, very tight. He was reading off a piece of paper! It was horrible."

Before the meeting broke up, Sigmund had announced the number of his hotel room in case anyone needed additional counseling. When the bar was closing that night, it was decided that the bill, a hefty one, fell within a broad definition of "additional counseling," and someone put Sigmund's room number on it. The next day it took hours to straighten out the charges.

Not everyone sought relief with alcohol. Kim Valentine had been released from Valley View Hospital in time to rejoin the hotshots for the meeting. "She was quite emotionally exhausted and shook up," said Connie Shaw, patient advocate at the hospital, where, as a polite convenience, the records say that Valentine was treated for "smoke inhalation." Yet Valentine proved herself an emotional anchor for the Prineville Hot Shots that night.

"Those boys drank their sorrows away," Valentine said. "I stayed up all night with those guys, just making sure they didn't jump off their balcony window. They kept asking, 'Why aren't you drinking?' and it's like, 'I don't need it.' I told them I have an advantage: I know about my faith. That night I think God allowed me to be calm for those guys."

The counseling story, fortunately, does not end with Sigmund's aborted session. In an act of enlightened generosity, the federal government subsequently paid for psychiatric therapy for its employees who could show that they had suffered emotional damage as a result of the fire. In several cases these sessions continued for years.

PRINEVILLE, POPULATION 6,100, is a hardscrabble remnant of Old West days, when cows and timber dominated the economy of central Oregon. Prineville had a heyday as a prosperous trading center, the principal one in the region, until the railroad passed it by in the early twentieth century. A plucky town, it refused to wither away; it built a city-owned railroad spur that remains in use as the City of Prineville Railway, said to be the only municipally owned railroad in the nation. With the cattle and logging industries in long-term decline, Prineville's largest employer today is the Les Schwab tire-distribution center.

After the first news flashes from Storm King, word spread through town that the Prineville Hot Shots had been on the fire. The hotshots were a local team, athletes who returned from fire season with action-packed stories and, in a welcome sign of maturity, their own money. They inspired the dreams of many a teenager; adults felt the same pride for them as for the Cowboys, the high-school football team. "They're our troopers, our military," said Dennis Olsen, a real-estate agent. "You read the papers about how the Prineville Hot Shots battled these fires, and it makes you awfully proud." Everyone knew the risks involved, but accepted them as part of Western living.

The list of dead and missing read out on the afternoon of the seventh contained the names of nine Prineville Hot Shots, five men and four women.

"They're not victims, they're casualties; it's like war," said Greg McClarren, a spokesman in Prineville for the Forest Service. "Nine of them gone. Eighty percent of the women, dead."

Flags across town went to half staff. Volunteers at the Prineville Teen Center made blue lapel ribbons to distribute. "They're going out of here quicker than we can make them," said Ginny Koehn, director of the center. "One death affects a whole community—but when you have nine, the impact . . . you can't describe it."

THE RESIDENTS OF Glenwood Springs, population 7,200, have spent more than a century making things easy for other people. Teddy Roosevelt used the Hotel Colorado as his Western White House on a presidential hunting trip. Buffalo Bill Cody bathed in the healing waters. Doc Holliday came to the town at the end of his career as a gunslinger and is buried on a hillside.

Today the town makes a living from steamy water, ice-cream parlors and yuppie restaurants. The residents have seen their share of real-life catastrophe, though, the kind that strikes home: In 1985 an explosion at the Rocky Mountain Natural Gas Company building in Glenwood Springs killed twelve; in 1981 a methane-gas explosion at a nearby coal mine killed fifteen, six of them Glenwood Springs residents.

This time, with the exception of Tyler, the victims were not neighbors, but it was still a hometown disaster. Purple ribbons became the symbol of public mourning. One resident, Jeff Feaster, tied strips of ribbon to light poles and handed out lapel ribbons to people he met. The color, he said, signified "the courage and honor of a fallen hero, like a Purple Heart."

The Glenwood Post's managing editor, Dennis Webb, caught the moment in a signed editorial: "Here in Glenwood Springs, a spa town with an emphasis on fun and recreation, the feeling today is far from idyllic. . . . These men and women took to the front lines to defend us as bravely as any soldier. They have forever earned our gratitude, we are forever in their debt."

IN THE SMALL hours of July 8, about 1:30 A.M., Pamela J. Miller awakened at her Glenwood Springs home, one the fire had threatened, to nurse her infant son. Miller had grown up in Prineville, graduating in 1983 from Crook County High School, where Jon Kelso was a fellow student, before moving to Colorado. In the dark and quiet she began to jot down an account of the past several days.

"Just over 24 hours before, we watched with horror and disbelief as the South Canyon fire reached the ridge of the hill behind our house," Miller wrote in a letter published in *The Glenwood Post*. "We tried to prioritize what we may need to take with us—clothing, pictures, documents? Because of the firefighters I am instead sharing this beautiful, sleepy, secure moment with my family in my house.

"In Prineville, where I grew up and where my family still resides, the town is small, the families are close, the children have all grown up together. I share in their grief.

"Out my window I can see the headlamps of the firefighters on the hill even now, working through the night to keep us safe. I pray, Heavenly Father, give them strength, keep them safe and bless them generously as they have truly been a blessing to us. Amen and Amen."

*   *   *

THE GARFIELD COUNTY sheriff set up a table at Canyon Creek Estates after dawn on July 8 to outfit and register volunteers for the continuing search for Tyler and Browning. The undersheriff, Levy Burris, was checking people in when a young man came to the table wearing a one-piece green fire suit, the kind used by helicopter crewmen. The man, who appeared drawn and tired, had a familiar face, though Burris had never seen him before.

Over the next several days Jim Tyler would cause more than one startled look as he was mistaken for his younger brother, Rich. Jim and his wife, Janis, had driven straight through to Colorado from their home in Milwaukee after hearing news reports that Rich had been in the South Canyon fire. They had arrived at Patty Tyler's house in Palisade, a few miles from Grand Junction, early on the morning of the eighth. The Tyler parents, Arthur and Ruth, were there, too. Jim and Janis had heard no further reports about Rich during their drive, and Patty brought them up to date.

While they talked, Winslow Robertson, who had been relieved as IC by the national team, showed up at the Tyler home. After some discussion he agreed to drive Jim Tyler to Storm King and try to place him on the search party. Jim borrowed a pair of Rich's heavy boots; Robertson gave him the flight suit and a pair of gloves. Tyler wondered, Why gloves? But he took them anyway.

"When we got to Canyon Creek Estates, Winslow told me to let him do the talking," Tyler said. "Everyone was in a kind of daze. My impression was Burris was embarrassed because they hadn't found them. They asked if I was a brother. They weren't sure what to do with me—so they let me go up."

When Tyler arrived on the ridgetop, he stepped into a layer of ash that rose and fell in waves. He felt heat through his boots and realized one reason he was wearing gloves: If he tripped or grabbed a root for support, he would be touching embers.

"I had the impression things had not gone right with the search, there had been mistakes," Tyler said. "But none of that mattered once I stepped out of the helicopter and looked around. The mountain was a smoldering cauldron. I realized at that point we weren't looking for someone who was alive."

Robertson, who had talked with survivors by then, passed on to Burris what he had learned about the last sightings of Tyler and Browning. It made no sense to search near their starting point, H2, Robertson said. Tyler had been a cross-country runner, "a real gazelle," and would have made it a long way.

The search party broke up into twos and threes. Robertson led Tyler along a path above the East Canyon, an area searchers had already covered. Within a few minutes Burris's voice came over the two-way radio.

"Is he with you?" Burris asked.

"Yes," Robertson replied.

"Can you get him back to the helicopter?"

"Did you find them?"

"Yes."

Jim Tyler let Robertson guide him to the helicopter. The emotional shock came on top of a lack of sleep and the change in altitude. "It was all I could do to get to the helicopter," Tyler said.

Tyler and Browning indeed had run a long way. They had begun their race with the fire by heading around the side of the knoll Shepard had used as an observation point. They could hear shouts and see hands waving as Shepard and Blanco directed them into the East Canyon.

The fire's natural path would carry it over the ridgetop and into the East Canyon. That was the last place they wanted to go. "Run the ridge, run the ridge," they called back.

The helicopter, 93 Romeo, hovered in full view above the East Canyon, but flames were too close for a rescue. If Tyler and Browning could find a high, open spot ahead of them, a pickup should be possible. Tyler had trained for such a moment two years earlier in the first helicopter rappel program offered in the Rocky Mountain region. When Tyler had briefed the new men, Browning and Steve Little, earlier in the season, he had passed along this lesson: When in trouble look for a high, open place. The helicopter crew will yank you out by your belts if necessary, he told them.

By the time Tyler and Browning skirted the knoll, the fire had crossed the ridge behind them, closing off the possibility of escape in that direction. The two helitacks neared the junction with the peak of Storm King with few choices left. Continuing straight up the peak was a job for mountaineers. Turning right would eventually take them into the East Canyon, which was filling with flames. About two hundred yards to the left, along the base of the peak and roughly on their level, a bare rock outcropping stood out like a thumb inviting a helicopter rescue.

They scrambled toward it.

They clattered through rockslides and dodged through oak brush. Game trails beckoned and then petered out. The wave of flame burned to the end of Hell's Gate Ridge, turned and began to follow them; at the same time the blowup in the western drainage roared toward the peak of Storm King. The two arms of fire formed a closing pincer, with Tyler and Browning in its grip.

They arrived at the foot of the outcropping, about a third of a mile from H2, and there met a surprise—a broad, eroded draw about fifty feet deep. They would have to crawl through the draw to ascend the outcropping.

The air heated up. The fire was closing in, and no helicopter was in sight. They clambered into the bottom of the draw. They had one option left, deploying fire shelters.

They dropped packs, flight helmets, a Pulaksi and other gear and pulled out shelters. Then they moved another thirty feet down the draw to a wide shelf, free of brush, a better place for a deployment. Browning was about eight feet below Tyler.

The fire struck with such force that it tore the shelters from their hands. It knocked both men flat on their backs, parallel to each other and crosswise to the draw, as though the arm of the fire from Hell's Gate Ridge had struck first. "It was as if they heard something coming, turned to face it and were knocked over backward," Putnam later observed.

They never moved again. Dust, stones and a few boulders loosened by the fire rattled down and partially covered them. Searchers in an aircraft could have spotted them, but only from directly overhead. The draw was so steep that a helicopter was called in to extract the bodies.

The discovery of Tyler and Browning closed the chapter on

the search-and-rescue mission forty-four hours—nearly two days—after the blowup.

BY THE END of that day, July 8, the effort to put out the South Canyon fire had turned the corner. "Team performing great," the IC, Jack Lee, noted in his journal. "Planning and briefings going well." Lee had about five hundred firefighters at his command, not an especially large number and not as many as he wanted—big fires often require several thousand firefighters. The weather helped by staying calm.

The media remained "out in force," Lee noted, but was finding less and less to report from the scene. His team would have a line around the fire, Lee predicted, by 6:00 P.M. the next day, Saturday, July 9.

He met that deadline, but it took an additional two days, until 6:00 P.M. on Monday, July 11, before the fire was declared controlled. The fire had been persistent, but it had grown little in the five days after the blowup; it never destroyed a man-made structure and never injured a civilian.

The cost of suppressing the fire came to $1,689,119—$957,228 for the BLM and $731,891 for the Forest Service, according to agency tallies.

The largest single bill was an additional $1,784,989 for payments to the families of the fourteen victims under the Public Safety Officer's Benefits Act of 1949, passed in the wake of the Mann Gulch fire. By 1994 the amount payable per victim had grown with inflation to $127,499.

Burial and related costs added another $53,905. By 1998, psychiatric care for BLM employees had added at least $15,631 more, and an ongoing study of the fire another $1 million.

Dating from the lightning strike on July 2, the South Canyon fire burned for ten days. It took fourteen lives, cost more than $4.5 million and destroyed 2,115 acres of commercially worthless scrub brush. By comparison, the average fire in Colorado over the previous five years cost $62,696 and killed no one.

Part

III

SIXTEEN

THE STRUGGLE OVER how the South Canyon fire would be remembered began long before the fire came under control. Who would be a hero? Who held responsible? What would be the lessons? The event included fourteen deaths, the largest number lost to wildfire since fifteen firefighters had been killed in the Rattlesnake fire in California in 1953, more than forty years earlier.

The echoes from Mann Gulch, the cries of "Let's get the hell out of here" ringing in another gulch ahead of another blowup, guaranteed Storm King a special place in history. Like those who fell in Mann Gulch, those who died on Storm King Mountain were the elite, the seemingly untouchable—not only smoke jumpers but hotshots and helitacks. And this time four were women, the first-ever instance of multiple deaths of women in wildland fire.

The role of Don Mackey became an enduring legacy of the fire. Mackey's fateful part in the decision to build a fire line downhill would be debated as strenuously as Wag Dodge's decision to light a separate fire and step into its ashes to escape the blowup in Mann Gulch; Mackey's decision to go back for others—not once but perhaps twice, as will become clear—would inspire future generations. Those actions cost him his life: What, then, were the consequences for Mackey's superiors, who had made mistakes but had been left untouched by flames?

Many of the choices that led to success and failure for the South Canyon Fire Investigation, the eventual report of Forest Service and BLM investigators, had been made by July 11, when the fire was declared controlled. Les Rosenkrance, the BLM's state director for Arizona who had been named head of the fire investigation, intended without fail to meet the forty-five-day deadline for reporting findings. The early deadline contributed to a decision to exclude from consideration events after the blowup on July 6 and to limit attention to events before that date. Sensibly enough, detailed recommendations for change were left to a separate group called the Interagency Management Review Team, which had a less hurried deadline.

"We were trying to get an awful lot done in a short time," Rosenkrance said. "We didn't try to go into things like, Why was the budget cut, Why was the head count reduced, those kinds of things. The fire started, the employees were killed, a little bit of what led to the [disaster], and that was the end of our report."

One early decision by Rosenkrance proved especially fateful: Since Dick Mangan and Ted Putnam were investigating the fire for the Missoula Technology and Development Center, Rosenkrance named them to his team to avoid overlapping efforts. From the first, Mangan and Putnam played the role of odd men out.

Within days Mangan felt the full burden of his past experience as administrative head of the Prineville Hot Shots. He held an emotional reunion with Tom Shepard at the Hilton Hotel in Grand Junction and then flew to Prineville to interview some of the Prineville Hot Shots. By the time he returned to Colorado, he had begun to wish he had not joined the investigation team in the first place.

But he also had strong doubts about the investigation itself. The forty-five-day deadline, he felt, allowed too little time to do justice to the hotshots. And Rosenkrance's team appeared to

Mangan to be bent on protecting managers at the expense of firefighters.

"I started getting this uncomfortable feeling," Mangan said later. "The emphasis was all on the mistakes the firefighters made."

Putnam had a similar feeling. He saw the unfolding inquiry as a rerun of the aftermath of the Battlement Creek fire—an exercise in management protecting its own.

"We compounded the Battlement Creek tragedy by ignoring the human failure issues that if addressed could have minimized future fatalities," Putnam complained in writing at one point. "We are making the same mistake in the South Canyon report."

The conflict over who should shoulder the most blame, fire managers or firefighters, eventually split the investigation team and was never resolved. The confusion angered and demoralized many of the victims' families and friends, and made it more difficult to learn the lessons of the fire.

The investigation got off to a bumpy start for other reasons as well. Rosenkrance watched unhappily as most survivors left Colorado before they could be adequately debriefed. He tried to stop the exodus, but the survivors had asked to go home and had won high-level support.

"People were leaving as we were arriving," Rosenkrance said. "We felt it was important to get their statements, their side of the story as quickly as we could, without having it changed as they talked to more and more people." The Forest Service ruled that its employees could go home, which meant that everyone based outside Colorado was free to leave the state—the BLM employees were local. The question of whether witness statements could or should have been compelled was never resolved.

On top of that, Rosenkrance's team had little experience conducting investigations. Mangan and Putnam had used interviews in assessing fatality sites, but had seldom taken statements themselves. Only one member of the team, the ranking Forest Service representative, Mark A. Reimers, deputy chief for congressional relations in Washington, had formal investigative training.

The team wound up taking notes by hand because no tape recorders were provided.

"Some of the people on the team did not take good notes," Rosenkrance said. "They wrote in their own very broken language, and if anybody else tried to read it, you couldn't figure out what they were talking about. We had to send them back several times to get some of that information.

"In contrast, when we interviewed some of the managers, I sat down and [wrote] down, what are some key points we need to know from this individual before we let him leave the room. We got just exactly the information we needed from some of those people.

"So we stumbled and we floundered around pretty rough."

A case in point was Kevin Erickson, who had argued with Mackey over the west-flank line and had been among the last to see the twelve firefighters before they were killed. Erickson was among the first of the survivors, if not *the* first, to leave Colorado. After his arms were bandaged at Valley View Hospital on the sixth, he spent the night at a motel in Glenwood Springs and went home to the Bitterroot Valley the next day, July 7.

Erickson was not interviewed until July 18, almost two weeks after the blowup, and then it was done by telephone conference call. He took the call in his bedroom. Nameless voices, talking over each other, peppered him with questions. No tape recording was made. Little wonder, then, that Erickson's story became mangled in the telling and subsequent reporting.

The most serious difficulty arose over Erickson's description of his two talks with Mackey, the first one when they had argued about the fire line and the later one when Mackey had offered, half jokingly, to let Erickson take over the fire. The official account of the interview mixes the two conversations together. Initial handwritten notes report: "He [Erickson] didn't like going down in there and told Macky [*sic*] so—Macky didn't really know who IC was him or Blanco since Blanco was so quiet."

Those notes were expanded, as should be expected, into a fuller version, and changed from a third-person "as told to" account to a more authoritative first-person or "I" point of view. Again the account failed to separate the two conversations.

The second version reads, "I didn't like going down in there. I talked to Mackey about it. Not burning too active. I voiced my opinion, I said that the stuff could go anytime. I was going by his judgment, his best judgment was to go direct. I did not like it. He offered to let me take over the fire. He said that he didn't know who the IC was, him or Blanco. He asked me if I wanted it. I said yes. He just smiled at me and said, 'We'll see what happens.' "

Mixing the two conversations together and leaving out the questions creates the impression that Erickson's disagreement with Mackey was far stronger than it actually was.

Erickson made matters worse for himself when he was given the interview to check. He crossed out several sentences critical of Mackey: "I voiced my opinion, I said that the stuff could go anytime," and, "I did not like it. He offered to let me take over the fire."

He then wrote in, after the remark that Mackey's best judgment was to go direct, a sentence helpful to Mackey: "I thought that was the best way also." It now appeared that Erickson had said too much to the investigators the first time around, casting his brother-in-law in a bad light, and then had tried to make up for it by altering his testimony. Or at least that was the interpretation given in newspaper stories when the interview was made public, with Erickson's handwritten changes, under the Freedom of Information Act.

Erickson said later that he was not trying to cover up anything. He made the changes, he said, for the sake of accuracy and to emphasize that in the end he had agreed with Mackey about the west-flank line. Erickson's explanation should be given weight, considering the confused circumstances of the interview, but unfortunately it was not included in media reports.

The Denver Post, which printed the first report, later published a letter from Erickson in which he complained, "Why must you continue to try to read more into our mistakes than what is actually there? Yes, we made mistakes, but Don was not alone when decisions were made. Many people will have to live with their decisions for the rest of their lives."

By the time the news reports appeared, Erickson had decided to leave smoke-jumping. Bob Mackey had counseled him after the fire to consider another occupation, given his family responsibilities. "I took that advice to heart," Erickson said. "You don't see one of your best friends killed and not think long and hard about it."

For several years the Missoula smoke jumpers kept Erickson's gear in a locker at the Aerial Fire Depot in case he came back. But Erickson found a job as a welding instructor at a Job Corps site in Darby. The work lacked excitement but provided year-round employment and allowed him to spend summers at home. He and Jan drew closer through the difficult times and became "best friends," in Jan's words, as well as husband and wife.

AFTER TELLING THE Mackeys the bad news about Don, Walt Smith stayed on in Hamilton to help them. Because Smith had been Don's friend and had rank in the fire world—he was fire officer in the Sula District in the Bitterroot Valley—he became the family's go-between with the Forest Service.

Things did not go well.

"There was sort of a meltdown" in the official handling of the fire's aftermath, Smith said later. "If I wanted something done, I had to call up; nobody called me with anything. The family was going through denial, grief, rage. I needed some support. But somebody'd call up and say, 'How's it going?' And I'd say, 'Not too goddamn good.' "

Matters came to a head over the issue of an autopsy. The Mackeys, who did not want an autopsy performed on Don's body, were told that their wish would be respected. Colorado law, however, requires an autopsy in cases of accidental death whether families agree or not, in part to determine if alcohol or drugs were involved.

Autopsies subsequently were performed on the fourteen bodies, but the Mackeys were not told of this until after the fact. Bob Mackey received a telephone call from Colorado with the news that Don's body, like that of the thirteen others, had been "clean" of any trace of drugs or alcohol.

"I could have told you that before you did it," Bob replied in outrage. "What about the people who didn't die, did you check them?"

★ ★ ★

MOST STORM KING survivors were interviewed in groups before leaving Colorado. The questioning was disorderly. Those who knew the most often kept silent; group pressure and time constraints tended to suppress detail.

The Prineville Hot Shots, true to their group culture, insisted on being interviewed together. They provided a bare-bones account and then left Colorado, which resulted in Mangan's being dispatched to Prineville for follow-up interviews.

"We wanted to get Prineville out of there as quietly as possible because the media was hounding them," said Mike Lowry. "We had two airplanes and basically snuck them out. We kept them hid for a while."

The Prineville Hot Shots dropped the name "Prineville" and joined crews needing replacements, disappearing from the public eye for the rest of the summer. Several who returned to work immediately were assigned to a fire engine in Prineville.

"We drove around the district for a couple of days and went to lunch, just to keep busy," said Alex Robertson, the Prineville sawyer. The others drifted back over the next weeks, with the exception of Bryan Scholz and Kim Valentine, both of whom decided never to return to the fire line.

Scholz became an assistant fire-management officer in Oregon. Valentine joined the ecology crew on the Ochoco National Forest in Prineville, working on projects such as aspen protection. "Everyone respected her decision not to come back," said Tom Shepard. "There's a lot of understanding." Shepard continued as Prineville's superintendent for one more fire season, 1995, to rebuild the crew, and then became an assistant fire-management officer in Idaho.

Public interest in the Prineville Hot Shots remained high. Late in the summer an enterprising reporter set out on foot after learning where Prineville was fighting a fire, far from any road, as part of another hotshot crew. The local radio dispatcher let the reporter walk in halfway before broadcasting the misinformation that the Prineville Hot Shots had been transferred miles away. The reporter turned back.

Most surviving smoke jumpers gathered in a motel room two days after the fire for their group interview. "I thought they would ask us some specific questions; instead it was this group thing, and some people never said a word," Sarah Doehring stated later.

Unhappy with the result, investigators asked the jumpers to write follow-up accounts. "End rewrite, got to catch a plane," Quentin Rhoades typed at the end of his ten-page report.

Doehring considered her written account no substitute for a thorough debriefing. "I think if they had taken more time and done a little better job of it, things would have been resolved more in people's minds," Doehring said. "A lot of things were left open. That's why a lot of people, families of people who died, have questions."

Doehring began to experience flashbacks and a strain on her marriage, symptoms familiar to combat veterans. The first night home in Florence, a small town in the Bitterroot Valley near Missoula, she awakened when a wind started up. Dark branches clawed at her bedroom window, and for a moment she was back on Storm King. Another night she awakened to the smell of smoke, but it was only a dream.

"For a long time, every day I'd think about Don. There's a car at the end of our road that's the same kind of car he used to drive to work. The fire affected me more than I realized; it made it hard on me at home."

The shared experience of catastrophe, though, brought new friendships.

"It's strange how I've created bonds with people I never would have otherwise, like Bryan Scholz," Doehring said. "When I saw him, he was really hurting; I mean, half his crew was wiped out. So I've become friends with him, and we stay in touch.

"It's not like I just got out of the Vietnam War, but I think about Vietnam and the guys who couldn't talk about what they saw or did, and that's how I feel about it sometimes."

The departure of survivors left BLM crew members to tell the story; many did not want to talk to outsiders. "We had a lot of pressure from the media and only two people talking," said Kathy Voth, a public-affairs specialist for the BLM.

Michelle Ryerson and Brad Haugh became spokespersons for the South Canyon fire. Ryerson escorted a succession of reporters up Storm King, her poised manner and "tall and slender" appearance winning positive coverage. One national magazine became so enamored of her that their lengthy account made Ryerson the incident commander, ignoring Butch Blanco, contrary to her standard account of the fire. Ryerson fought fires the rest of the summer, to show she had not been intimidated, and then became a radio dispatcher. In time her voice sounded over the loudspeaker outside the smoke-jumper shack at Walker Field.

Brad Haugh's heart-pounding story of his race up the west-flank line—including the made-for-headlines phrase "Hell Caught Up with Us"—became a staple of media accounts. His story, though, had a major defect: his claim that he stood face to face with Jim Thrash—"As close as I am to you now!"—and heard the smoke jumper say "Deploy? Deploy?" Disturbingly, Haugh's story left the impression that Thrash and perhaps others with him could have survived if they had followed Haugh.

At least on the surface, Haugh's story failed to square with the accounts of Kevin Erickson and Eric Hipke, both of whom had been there. When Haugh persisted, Erickson and Hipke met for dinner in Missoula, without Haugh, to compare their stories point by point.

The degree of injuries makes it certain that the three men escaped over Hell's Gate Ridge in this order: Haugh first, sustaining a slight burn on the elbow; Erickson second, with second-degree burns on both arms; and Hipke last, with burns over 10 percent of his body.

Erickson remained at The Tree until after Haugh left, but Erickson never saw Thrash up close. Erickson did see a bearded firefighter well below him, too far away to tell if it was Thrash or Roth.

Hipke, coming up from below, passed The Tree moments after Erickson and Haugh had left. Hipke had just looked back and seen Roth and the others well below him.

For Haugh's story to be true, then, Thrash would have had to be superhuman, spurting past Hipke without being seen by him, coming face to face with Haugh without being seen by Erickson, who was next to Haugh, and then disappearing as Hipke passed The Tree.

Erickson, Hipke and members of Rosenkrance's team who looked into the matter concluded that Haugh suffered from an overly vivid imagination. "Haugh would have been one of the heroes of the fire," Erickson said, "if he hadn't claimed he saw Thrash."

Haugh, though, did hit upon what may be the most bizarre link between the Mann Gulch and South Canyon fires. The following diagram, put together by Haugh, uses the numerals representing the dates of the fires—August 5, 1949, for Mann Gulch and July 6, 1994, for South Canyon:

$$8—5—49$$
$$7—6—94$$

There is enough coincidence in the numerology to give a skeptic goose bumps. A simple switch of numbers ties the years together: 49 for Mann Gulch and the opposite, 94, for South Canyon. The numbers for the days and months, excluding the years, proceed in a mathematically perfect circle, starting with 5, the day of the Mann Gulch fire, then continuing clockwise to 6, the day of the South Canyon fire, and then to 7 and 8, the months of the fires.

Matters become spookier when the numbers are added up. The dates for the years, 49 and 94, add up to unlucky 13 each. The dates for the months and days of each fire, 8 and 5 for Mann Gulch and 7 and 6 for South Canyon, also add up to 13 each. Thirteen is the number of people killed in Mann Gulch. (The 14 killed at South Canyon could be reduced to 13 by subtracting Mackey, who made a conscious choice to return along the west-flank line, but that is stretching a point, and Haugh did not suggest it.)

Haugh became obsessed by the fire. He had his left shoulder tattooed with the Prineville dancing coyote and nine stars, one for each Prineville victim, and planned another tattoo to commemorate the three smoke jumpers. He began to carry a disposable camera to document his observations on fires. His behavior became so erratic that, by his account, he was relieved from fire-fighting duty after threatening a fellow employee, but with the understanding he could return after counseling.

UPON RETURNING TO Colorado from Prineville, Mangan began digging into the performance of BLM managers in Grand Junction. The evidence was overwhelming that the BLM's Grand Junction District had failed to follow its own policies requiring immediate suppression of fires. It appeared to Mangan that the district had conducted "business as usual" despite the drought.

"They [the district] weren't thinking big picture," Mangan said later. "They were a Class-B, six-person football team suddenly playing in a stadium with ten thousand people, and they didn't know what to do."

Asked why they had not responded immediately to the South Canyon fire, BLM managers repeated the two stock answers: There were too many fires—"all resources were committed"— and Storm King was "inaccessible."

When Winslow Robertson told Mangan during an interview how he would "bop on over" to the Western Slope Coordination Center to see what airplanes were available, Mangan became indignant. He asked Robertson if the Western Slope center did not cover districts as far away as Moab, Utah.

"What do folks in Moab, Utah, do?" Mangan asked Robertson. "Do they 'bop on over' to Grand Junction and see what's on the tarmac and then drive back home and cut the resource order? Don't you call up and say, 'I need one, get it for me, and if you don't have it, go to the next higher level, and if that doesn't work, go to Boise'?"

As the scope of the Western Slope–Grand Junction District feud began to emerge, the BLM's state director, Bob Moore, was called on by Rosenkrance to explain why he had not settled it.

"He [Moore] said, 'It's been going on forever, so why are you people trying to blame me for it?' " said Rosenkrance. Moore also blamed budget and personnel cutbacks for adding to tensions, which, he said, led to a reluctance by managers in Colorado "to support aggressive initial attack on fires."

Rosenkrance eventually concluded that the Western Slope–

Grand Junction dispute had not caused the fire deaths, but it "sure as hell didn't help either." Rosenkrance was no stranger to the controversy; he had been one of the auditors in 1989 who found "serious morale problems" in the BLM's Grand Junction District.

Mangan began hearing a rumor that retardant tankers had been available for Storm King but had not been used. When he checked air logs, he discovered that tankers indeed had been ready at Walker Field morning after morning.

"They bet on the come, they bet they were going to have a worse fire start in the afternoon than the ones they already had," Mangan said later. He also found gaps in logs, when conversations had occurred but not been recorded, a serious breach of procedure. And he discovered that Blanco had used a cellular telephone on Storm King, but none of those calls had been logged either.

The findings fueled Mangan's suspicion that the BLM's Grand Junction District was an "old boy" network dominated by cozy personal relationships. Perhaps, he thought, action on the South Canyon fire had been deliberately delayed to "save" the fire for Blanco, the local hero. This charge was roundly denied by BLM supervisors, and no evidence ever turned up to support it.

But it began to look to Mangan as though a member of the investigation team, Roy Johnson, who once had been fire-management officer for the Grand Junction District, was taking the district's side in every dispute. Johnson later scoffed at the notion that he was biased.

"I don't think Mangan or I had any impact as a result of our past experiences," Johnson said later, adding that he gladly would have skipped being an investigator.

"There were people who wanted to hug you, but you're part of the investigative team. You want to heal with them, but the healing process hasn't started for you." Johnson welcomed a suggestion that future investigation teams be balanced with members from outside the ranks of firefighters.

Mangan asked for a meeting with Blume, Robertson and Paul Hefner, director of the Western Slope center, to go over the evidence he had accumulated. Rosenkrance called the meeting for July 21, the day before his team was due to leave Colorado.

Mangan laid out his findings. When he asked why available

tankers had not been assigned to Storm King, Blume and Robertson responded that there had been too many other fires with higher priority.

If that was true, Mangan asked, why had they failed to order more tankers and other resources from outside the region? He made a note of Blume's response: "Don't ask for resources you don't think you'll get." Mangan then turned to the director of the Western Slope center, Paul Hefner.

"I said, 'If at nine o'clock in the morning on July third, fourth, or fifth, the Grand Junction District had called for those [retardant tankers] to be assigned to the South Canyon fire, would they have gotten those resources?'

"And Paul said simply, 'Yes.' "

Rosenkrance by this time was shaking his head. Mangan took that to mean that Rosenkrance at last understood how badly fouled-up the management situation had been. That assumption was correct, but only up to a point. "Dick was absolutely right" about air tankers being available but not used, Rosenkrance said later.

Furthermore, Rosenkrance had uncovered on his own the BLM policy in Colorado of not using tankers unless firefighters were on the ground to follow up. "It was just flat-ass dumb," Rosenkrance said. He was so outraged that he told Bob Moore, the BLM's state director, to replace Don Lotvedt, the BLM's state fire-management officer, who lacked fire experience. Moore resisted the move.

"I said [to Moore], 'Right now I can't say you are responsible for those fourteen people who got killed,' " Rosenkrance later recalled. " 'But I can say you're accountable. I'm also telling you you're going to kill someone else the way this operation is going, and if you do not get somebody here [to replace Lotvedt], you will be responsible *and* accountable. And I'll by God see that you are!' "

Lotvedt was moved to a different job and subsequently took early retirement at his own initiative. Moore, who said at the time that moving Lotvedt had "no causal relationship" to the South Canyon fire, retired shortly thereafter. Asked if Moore's retirement was forced or voluntary, Rosenkrance replied, "It was forced." Rosenkrance said that after higher-ranking BLM officials apparently used too gentle an approach in urging Moore to retire, Moore asked Rosenkrance to clarify for him what was going on.

"I told him that he needed to pull the plug and get out of the organization," Rosenkrance said. "I encouraged him to do it." Moore declined to be interviewed for this book.

Rosenkrance, however, did not agree with Mangan's view that BLM managers bore the heaviest responsibility for the fourteen deaths. In Rosenkrance's judgment, no upper-management mistake had been significant enough to cause the fatalities. But Don Mackey's mistakes were another matter.

"When Don Mackey decided they were going to build that fire line downhill, that got them committed," Rosenkrance said later. "Really, if you had changed most of the other [mistakes], you still could have avoided what took place.

"Being first jumper out the door decides who is in charge? It's ridiculous. How in hell Mackey was in charge is beyond me." Rosenkrance had been a smoke jumper himself for eight years, and being first out the door during that time meant supervising only if you had the most experience.

"When I was a four-year jumper, I was invincible. You get a bad fire, and hell, let's get down where it's hot and stop that sucker. As I got beat on and crippled up, I started gaining a little more respect. I don't think Don Mackey had achieved that level yet.

"I think he was a fine man, but I don't think he should have been in charge."

The picture of Mackey painted by Rosenkrance is one of an overly aggressive, woefully inexperienced smoke jumper who failed to rise to a challenge and who virtually alone made the decision to build the fire line downhill. But that view of Mackey does not fit the facts as they are now known.

Rosenkrance relied on the results of his team's inquiries, and so was unaware of the full extent of Mackey's doubts about the fire—his worried vigil at the campsite the night before the blowup, his repeated remark that there was nothing on Storm King worth getting hurt over, his attempt to pull back when a tree torched on the west-flank line, the details of his two exchanges with Kevin Erickson. Nor was Rosenkrance aware of the shared decision making about the west-flank line among Mackey, Butch Blanco and, to a lesser degree, Rich Tyler. Blanco's description of the process—"It was a cooperative"—was given in an interview during a separate investigation by the Department of Labor.

Rosenkrance's investigation uncovered none of Mackey's last three radio transmissions—his statement to Dale Longanecker about going back to check on the "people in the brush patch," the order overheard by Archuleta to clear the west-flank line and the final transmission heard by Tom Shepard: "We're coming up to a good spot." Those quotations and much of the fresh information, painting a more complex picture of Mackey and of events, came to light during inquiries for this book.

In a conversation a few years later, Rosenkrance seemed to have mellowed a bit on Mackey. "He did what I might have done at that stage in my career," Rosenkrance said. "He tried to do the right thing." Mackey was aggressive—by nature, training and circumstance. He had been thrust into a supervisory role on his first fire of 1994, his first as a career smoke jumper. That would give anyone an adrenaline rush.

But he did not ask for the job, in fact had never been interested enough in supervision to take a single qualifying course. If Mackey had lived, he likely would have gone on to be a jumper instructor, "not sitting in some office in a Smokey Bear uniform," said his father, Bob.

For a man not given to doubts, Mackey expressed a remarkable number of reservations about the South Canyon fire, but he kept being drawn in by circumstances beyond his control. His doubts failed to overcome the cumulative effect of a long list of mistakes, most not of his making: The fire should have been put out days earlier, reinforcements should have been sent sooner, and Mackey should have been relieved as smoke jumper in charge. Those errors combined with the fire's deceptive behavior to cause disaster and cost Mackey his life. Other responsible parties, as will be seen, came to happier ends.

As the summer of '94 drew to a close, heavy thunderstorms swept western Colorado, ending the drought.

At about 10:15 P.M. on September 1, a motorist driving through heavy rain on I-70 past the foot of Storm King heard "a whoosh like a real strong wind going through the mountains." Hundreds of tons of mud, blackened trees and scorched brush, loosened as a result of the fire, slid down gullies, spilled across I-70 and poured into the Colorado River. The mud engulfed thirty vehicles. Traffic on I-70 backed up for four miles.

Several people and vehicles were swept into the river. Two people were injured, but no one was killed.

SEVENTEEN

ANGAN AND PUTNAM headed home to Missoula
with the conviction that the South Canyon fire investigation was
a deeply, perhaps fatally, flawed exercise. The forty-five-day
deadline was too short; team members lacked objectivity; interviews had been botched. Worst of all, as far as Putnam was concerned, too little consideration was being given "human factors,"
the decision making in Grand Junction and on Storm King
Mountain.

The investigation team aimed for an August 22 press conference to announce its findings. The occasion would offer the best
opportunity to put the disaster to rest in the public mind. None
of the many workshops, seminars and investigations that followed
would command comparable attention.

A public hearing after the Mann Gulch fire had played a similar
role and at least in one way, by questioning witnesses in open
session, had proved more satisfying. A board of review, meeting
in Missoula September 26–28, 1949, heard testimony from witnesses, including managers up to the rank of forest supervisor; the
survivors, Wag Dodge, Walt Rumsey and Bob Sallee; and the

most outspoken critic of the Forest Service, Henry Thol, Sr., a retired Forest Service ranger and the father of Henry Thol, Jr., killed in Mann Gulch. The transcript of Thol Sr.'s remarks, a mix of incoherent rage and insider knowledge, makes painful reading even a half century later.

Upon arrival in Missoula, Putnam began sending in suggested changes for the report being pulled together in Rosenkrance's home base, Phoenix, Arizona. "When we came away from Colorado, we were pretty much of one mind how everything took place," Rosenkrance said later, at a time when he had become the BLM's national director at NIFC for fire and aviation. "Sometimes it's harder to figure out why."

Putnam began to feel, though, that his comments were not being taken seriously. "Whenever I brought human factors up, it was treated as though it wasn't real data," Putnam said later.

Putnam disputed, for example, the repeated description in the working draft of survivors "running" up the side of Hell's Gate Ridge, a more significant debate than it first appears to be. Putnam as a younger man had competed in cross-country races of fifty to a hundred miles and knew from experience that the incline on Hell's Gate Ridge where the twelve firefighters had died was too steep to be run.

The word "run" implies that the victims made a maximum effort to escape. Putnam calculated, to the contrary, that the twelve firefighters could have moved as much as 15 percent faster if they had thrown away tools and packs immediately. With that time advantage everyone on the west-flank line could have made it to safety. In Putnam's view, the report was about to miss an opportunity to highlight the dangerously close bond between firefighters and their tools and teach a safety lesson.

Putnam requested that "walked" be substituted for "ran," but lost the battle.

As the August 22 deadline drew near, Putnam and Mangan prepared to travel to Phoenix for a wrap-up session, but at the last minute Mangan was called away to be operations chief for a fire near Missoula. Putnam wound up going to Phoenix without him.

When he arrived in Arizona, Putnam found few if any of his

suggestions incorporated into the draft of the report. "Nothing had changed," Putnam said later. "Most of my stuff was saying, 'Hey, let's be a little tougher on the BLM district.' "

Putnam turned in his proposals again, but none made the next draft of the report. When he asked the person responsible for editing the draft what had happened, he was told that his list had been lost. So he turned in yet another version of his proposals, this time with the word "SAVE" written in red letters at the top of the page. That version, too, was lost, Putnam was told, fueling his suspicion that a general effort was under way to downplay mistakes made by the BLM's Grand Junction District.

On his last day in Phoenix, Putnam was chatting with Roy Johnson, the former BLM fire-management officer in Grand Junction who also was on the investigation team, when the conversation turned to Putnam's old nemesis, the Battlement Creek fire.

Johnson mentioned that he had seen Leonard Coleman, Putnam's rival sector boss, supervising a fire crew the first season after Battlement Creek. The remark stopped Putnam in his tracks. Putnam had kept his mouth shut at the time because he had been promised that Coleman would never fight fire again.

"That was the point I decided that if anything was left out, I wouldn't sign the South Canyon report," Putnam said.

(Coleman's return to firefighting was brief. Unable to put the Battlement Creek fire behind him, he soon asked to be relieved from firefighting duties. "I don't want to use the word 'haunted,' " Coleman said years later. "I had done the best I could. But anytime you have people killed working for you, you wonder if you could have done something different.")

Putnam returned to Missoula with the draft report and a burning sense of betrayal. He drove to Mangan's fire camp forty miles outside Missoula near the Blackfoot River.

Mangan agreed that the report put too much blame on firefighters and not enough on BLM managers in Grand Junction. He and Putnam felt they could not sign the document as it stood. Putnam sent a computer e-mail message to team leaders on August 15, saying he had lost confidence in the report. "I would like my signature removed from the signature list," Putnam said.

Mangan faxed several pages of suggested changes from his Blackfoot River fire camp and soon thereafter was contacted by the team's ranking Forest Service member, Mark Reimers. "We had five phone conversations where he called me in fire camp," Mangan said. "There was an increasing amount of pressure. This was happening on a Wednesday, and the intention was to have the news conference on Monday, nationwide—and two of us had refused to sign."

Mangan and Putnam say team leaders never threatened them. On the contrary, when they maintained their refusal to sign, Reimers told them to write separate letters explaining themselves to the Forest Service chief, Jack Ward Thomas. If Thomas agreed with them, they were told, the August 22 news conference would be canceled; if he did not, they could attach a minority report to the main one.

Mangan in his letter noted that the report was "thorough and well-written," and acknowledged that firefighters had made "significant errors." But if the BLM's Grand Junction District had followed their own policies, he said, the fire would never have spread beyond ten acres.

Of particular concern to Mangan was the failure to brief the Prineville Hot Shots about local fuels and weather and to pass along Red Flag Warnings. "The report, while acknowledging this failure, dilutes it by referring to the on-the-ground firefighters' failure to ask for a spot weather forecast," Mangan wrote Thomas. "I cannot ever remember asking, or being asked, for a spot weather forecast on an initial attack. I am confident that, had a Red Flag Warning been broadcast, the smoke jumpers and Prineville would not have been on the fire's west flank."

After receiving that letter, Reimers asked Mangan exactly what changes he required to sign the report. Mangan replied that the BLM's failure to brief Prineville had to be raised from a "contributory" cause of the deaths to a "direct" cause. Reimers obtained Rosenkrance's agreement to the change, and Mangan signed the report.

Putnam, in his letter, asked for a delay, one more meeting, to hash out his criticisms, but was told there was no time left. Putnam then wrote a second and final letter to Thomas, dated August 18:

"My responsibility was to answer the question, 'Why did they die?' Because key evidence is still unavailable—coroner's evidence and critical witness statements—my technical report remains incomplete and this question unanswered. . . .

"From the beginning, this investigation was a political and media event that compromised the team's effectiveness and led to an unrealistic completion date for the report. . . . I do not feel this report is of the technical quality it should be to withstand the public scrutiny that will follow its release."

Reimers contacted Putnam the next day, August 19, in a last attempt at conciliation.

"I took care of many of the concerns you raised with me—I know it is not your complete list," Reimers told Putnam in an e-mail message. "I believe you are generally supportive of the report. I recognize your objection to the 45 day push."

Reimers told Putnam he would try to provide him with the latest draft of the report, but after that, Reimers said, "It is your choice."

Reimers, contacted in retirement, said the above account of his actions is accurate. "My objective was consensus," Reimers said. "But I realized Ted Putnam was a researcher and probably could not in good conscience sign the report, and would go on doing research on this, perhaps for years. I told [Rosenkrance and Thomas] it was unlikely he was going to sign, and we shouldn't worry about it."

Putnam felt that too many issues had fallen by the wayside, and he maintained his refusal to sign. A telephone conference call then was arranged among Mangan, Putnam and Rosenkrance. The discussion became heated. "Les kind of lost it a bit," Mangan recounted, "and said to Ted, 'Goddamn it, you had a chance to make a change, and you didn't do it.'"

Putnam lost his composure. "Goddamn it, I tried to make those changes, and you goddamn people wouldn't do it, and I'm not signing your goddamn report."

Mangan leaned over the conference telephone and said, "Les, I think that's the end of the conversation."

Putnam never signed the report.

Rosenkrance felt frustrated and angry. "Just very candidly, I'm really pissed at them," he said later. "I don't understand their complaint. I understand Ted Putnam is a doctor [doctor of philosophy] and is going to take years to study something infinitum, and that's fine.

"In Phoenix I put my arm around him and said, 'Is there anything that bothers you, Ted, do you feel okay with all this?' And he said, 'Yeah, if this stays the same, I don't have any problem signing. But I've had them change these on me after I've signed, and I don't want to sign until later.' "

Rosenkrance said that Mangan's ties to the Prineville Hot Shots and Putnam's to the Battlement Creek fire clouded their judgment. The same might be said of Roy Johnson's tie to the BLM's Grand Junction District, Rosenkrance acknowledged, but mainly he was upset with Mangan and Putnam.

"I should not have invited them on the team," Rosenkrance said. "They got so defensive of the people who died that they could not say that those people were at fault in any way.

"And they were. They made a bunch of mistakes.

"If you were one of those young hotshots, this is your first year of fire, you had a very low level of responsibility. But as you look at the people who were there, the amount of training and experience, they needed to be speaking up louder. They had a responsibility to those people who didn't know what was going on, to make sure they were out of harm's way.

"And they didn't do that."

Three days before the news conference, the NBC television network reported that investigators would blame the disaster on "overzealous" behavior by firefighters. Though the word "overzealous" does not appear in the South Canyon Fire Investigation, the NBC report set a tone. "After NBC did that story, I was afraid they were going to try to blame my sister for doing her job," said Alisha Bickett, Tami's sister.

When the South Canyon Fire Investigation was unveiled on Monday, August 22, it blamed no individual. Instead it listed a bewildering 209 "findings" of cause, ranked according to whether they had "significantly contributed," "influenced" or "did not contribute" to the disaster. The attention given each finding was so slight that the overall effect was of touching every base without thoroughly examining any one.

The most controversial finding was a single sentence that blamed the "can-do" attitude of the smoke jumpers and hotshots for leading to a compromise of basic safety rules. That charge, the apparent basis for the NBC report, caused anguish for families, who saw it as blaming their loved ones for their own deaths, and outrage in the fire community, which *prides* itself on a "can-do" attitude.

"It's pretty convenient to blame people who can't defend themselves," said Bob Mackey in remarks reflecting a common complaint. "What I want to know is why these people who are supposed to be finding out what really went wrong up there don't have a 'can do' attitude."

The phrase had been meant to describe an "overly" aggressive attitude, Rosenkrance said two years later, when ill feeling continued to run high. "We should have explained that better—you want crews to have a 'can do' attitude."

The press conference, held in Denver, gave the media an opportunity to discover for themselves the high price of attempting to fix blame. One reporter tried to have it both ways, focusing blame on a single person while dodging responsibility for having done so.

"Without putting the blame on any one person," the reporter asked, disingenuously, "what one person could have made a decision that would have stopped this from happening?"

Mark Reimers easily ducked the question, saying that it would be "unfair" to focus on any one person. But the question achieved fame in the fire community, where it came to be cited as a moral low point in the "blame game."

The South Canyon Fire Investigation placed much responsibility for the disaster on natural causes—fuels, weather and topography. Gambel oak, it was noted, had been recognized as a treacherous fuel as early as 1976 in the Battlement Creek fire report. A member of that investigation team, Robert Mutch, had prepared a special alert about the danger.

Gambel oak normally is difficult to burn, Mutch wrote in 1976, but when dry "becomes one of Colorado's most flammable fuels." On the Battlement Creek fire an unseasonable frost had turned oak leaves brown. On Storm King the dryness had been caused by drought, and the oak leaves remained a misleading green. But the effect was the same, a dangerously low level of plant moisture.

Mutch cited a 1974 report by the Colorado State Forest Service that had warned of a "fuel type X," desert brush including Gambel oak, which burned readily when dry. "This report should be required reading for all fire management agencies in Colorado," Mutch wrote. By 1994 the lesson had been forgotten.

"So often we don't have the fortitude to go forward and seek the changes we need," Mutch said after the South Canyon fire. "I don't know that our report on Battlement Creek made any difference."

The South Canyon Fire Investigation also cited as causes of the disaster numerous violations of the ten Standard Fire Orders and eighteen Watch Out Situations, the basic guidelines for fighting fire. The standard orders were formulated by a Forest Service task force in 1957, after studying Mann Gulch and a handful of other major "tragedy" fires. The eighteen cautions were developed later.

The standard order most directly related to Mann Gulch is number five: Know what your fire is doing at all times. On Storm King Mountain as in Mann Gulch rule number five was at the core of the disaster: No one recognized the developing blowup and acted on that knowledge until it was too late, though the topography was almost identical, a funnel-shaped gulch with the narrow end opening at a major river, the Missouri River in the case of Mann Gulch. After the fire, the New Castle Volunteer Fire Department pieced together aerial photographs of the scenes in Montana and Colorado; the result could be twin halves of the image in a Rorschach inkblot test.

The 1957 task force also found a pattern common to fires involving fatalities, and not surprisingly it fits Storm King. The fatal pattern includes a complacent attitude on the part of fire personnel, lulled by easy fire seasons, which well describes the BLM's Grand Junction operation in 1994.

Wind and weather always make a contribution. Conditions preceding the "tragedy fires" were usually so hot and dry "that only a change in wind was needed to make the fire situation extremely critical." That perfectly describes conditions on Storm King Mountain on July 6, 1994.

A final element, "lack of information and experience in the potential of fire under extreme burning conditions," translated on Storm King into a failure to use available knowledge about the burning potential of Gambel oak.

But drought, wind, fatigue and inexperience define any busy fire season. Firefighters agree that strict adherence to the standard orders and guidelines would mean letting many, perhaps most, fires burn unchecked.

The South Canyon Fire Investigation, however, cites violations of a substantial majority of the rules—eight of the ten Standard Orders and twelve of the eighteen Watch Out alerts. The commentary about each violation is brief but to the point, and is given here in its entirety for the standard orders:

10 STANDARD FIRE ORDERS

1. *Fight fire aggressively but provide for safety first.*
   —The tactics as implemented provided for aggressive suppression but overlooked many critical safety factors.
2. *Initiate all action in response to current and expected fire behavior.*
   —Aggressive attack continued in spite of onsite indicators of extreme fire behavior and increasingly stronger winds. Most firefighters were unaware of or disregarded how intensely Gambel oak and pinyon-juniper fuel types burn during extremely dry and windy conditions.
3. *Recognize current weather conditions and obtain forecasts.*
   —No spot weather forecasts were requested by fire personnel. No onsite weather observations were taken. The investigation team could find no one on the fire who knew of the Red Flag Warning predicted to accompany the cold front.
4. *Ensure that instructions are given and understood.*
   —Instructions appeared to be fairly straightforward.

5. *Obtain current information on fire status.*
—No one on the fire had a complete picture of the fire's activity and status.
6. *Remain in communication with crew members, your supervisor, and adjoining forces.*
—Radio communications were good.
7. *Determine safety zones and escape routes.*
—Most of the firefighters did not have clear instructions on safety zones and escape routes.
8. *Establish lookouts in potentially hazardous situations.*
—No one could see the part of the fire that presented the most hazard.
9. *Retain control at all times.*
—During the first phases of the fire, supervisors effectively controlled the firefighters. Supervisory control was generally effective given the blowup conditions.
10. *Stay alert, keep calm, think clearly, act decisively.*
—The firefighters were alert, but they failed to adjust strategy and tactics in a timely manner. The firefighters remained calm during the events leading to the blowup. Failure to recognize the indicators of blowup conditions led to the entrapment of the firefighters. Decisive action resulted in the escape of 35 firefighters when the fire blew up.

The Watch Out Situations contain several specific warnings about the dangers of constructing fire line down a hill. The South Canyon Fire Investigation concludes that firefighters failed to follow those guidelines on the west-flank line, especially one that says that a crew should be able reach a zone of safety from any point on a downhill line "if the fire unexpectedly crosses below them."

The main message of the South Canyon Fire Investigation is this: Relearn the old lessons. Firefighters do not need new training, but instead should go back to the basics. "We believe that training is not the core issue," said a letter accompanying the South Canyon Fire Investigation. "Rather it is one of implementing the training all firefighters receive."

The follow-on Interagency Management Review Team came

to the same conclusion on June 26, 1995, nearly a year after the fire. No "novel approach" exists to shield firefighters from harm, the team declared. The review team proposed thirty-five measures to strengthen existing safety measures, but acknowledged that the list contained "no dramatic changes."

AFTER THE AUGUST 22 press conference, copies of the background materials for the investigation were sent to the fourteen families, whether they asked for them or not. The Mackeys became armchair investigators. Bob Mackey compiled a list of twenty-eight items of alleged wrongdoing based on the official data, but the material could prove misleading. Item number four on his list, for example, repeats the accusation that the New Castle Volunteer Fire Department "offered to send crew to extinguish fire," but "were denied by BLM," when in fact no such offer or refusal occurred.

The BLM and Forest Service passed along other material, ranging from sympathy letters to autopsy reports. When Don's autopsy report showed up, unwanted, Bob intercepted it before Nadine could see it.

The Mackeys began to look for a lawyer.

MANY CHANGES CAME to the fire world with no prodding. Within months of the fire, smoke jumpers on their own initiative abandoned the practice of first jumper out the door automatically becoming smoke jumper in charge. Each jumper, not just every other jumper or so, was now issued a radio.

The Western Slope Coordination Center was disbanded and its duties assigned to the Rocky Mountain Coordination Center. The BLM's Grand Junction District updated the radio equipment in its dispatch center. The National Weather Service was asked to make a sharper distinction between warnings for lightning and for wind.

Slogans appeared. Chief Thomas of the Forest Service called for a "passion for safety," and the secretaries of the departments of agriculture and interior announced a policy of "zero tolerance" for careless and unsafe actions. Thomas ordered a further study of the fire by a private company, Tri-Data Corporation of Arlington, Virginia. That report, issued in 1998, contained so many broad goals—eighty-two, including subpoints—that a further study was necessary to trim the list. Tri-Data concluded, among other things, that experienced personnel should be retained with incentives such as better pay, not pushed into early retirement.

The most revolutionary change came from firefighters themselves. Crews in the summer of '94 began to "just say no," refusing to build fire lines downhill no matter what safety measures were taken. In September, when a fire burned several square miles in the Clearwater National Forest, a downhill fire line was ordered.

"We took every precaution. We set lookouts, we set a time for pulling back. It was as safe as it could be—and the crew still refused to build downhill fire line," said the operations chief, Roger Jansson, son of John Robert Jansson, the ranger in charge of the Mann Gulch fire.

Refusals became less common with time, but they posed a lasting challenge to old, authoritarian ways. In the past, firefighters refusing dangerous assignments had found themselves verbally abused and subject to formal sanctions, including being fired. The debate on this issue continues today.

The most politically charged issue surrounding the fire concerned the role of women. Neither the South Canyon Fire Investigation nor any subsequent official inquiry touches this point.

The South Canyon fire was no coming-of-age event for women, who by 1994 had shared the same risks as men on fire lines for three decades. During those years three women had been killed in separate blazes in Montana, Idaho and Arizona. But the high number and proportion of women killed on Storm King was staggering: Of the forty-nine people on the fire, six, or 12 percent, were women, but a much higher percentage—four of fourteen, or almost 30 percent—were among the dead.

As a consequence women came to have a new awareness of the possibly brutal outcome of fire work, a realization not limited by gender. "Firefighting is like combat . . . except you're not supposed to die," said Bryan Scholz, who, like Kim Valentine and Michelle Ryerson, left the fire line for good.

The public response from families of the Storm King women

came quickly and was limited to mourning them as individuals and firefighters, not as special cases.

"She wanted to be a firefighter from the time she was seventeen," Ralph Holtby, the former smoke jumper, said of his daughter, Bonnie, who had joined the Prineville Hot Shots when she was eighteen. "I encouraged her. Sooner or later something like this had to happen; you never think it's going to happen to you, to your family."

In private a few family members of male victims expressed long-standing opposition to women in fire work. An occasional letter to the editor echoed that view. But overall the opposition was fragmentary and caused no groundswell.

It did raise a nagging point, though, one that should have been addressed in an official inquiry. Had the natural concern of men for the welfare of women slowed down the group on the west-flank line and thus contributed to the deaths of the twelve firefighters there? There were accusations to that effect after the fire, and gender equality has broken down on this point before.

In the Arab-Israeli war of 1948, the hard-pressed Israeli Army assigned women to combat. But the practice was dropped when the Israelis discovered that it reduced unit effectiveness, both because men showed disproportionate concern for women soldiers and because Arab units fought on rather than surrender to women.

The person best able to say what had happened on the west-flank line was Eric Hipke, who had escaped steps ahead of the twelve. Hipke had been interviewed on July 9, three days after the fire, but the session had been brief because of his injuries. It took many interviews and return trips to Storm King for him to reconstruct the retreat along the west-flank line.

As the twelve started back, the two Prineville rookies, Terri Hagen and Kathi Beck, were right behind Hipke, who was third in line behind Jim Thrash and Roger Roth. The position of Hagen and Beck, whether by design or luck, was the proper one for rookies—sandwiched between experienced personnel. Their location also meant that Hipke acted as a buffer between them and Thrash and Roth, reducing chances that the leaders slowed the pace out of concern for Hagen and Beck.

The two other women, Tami Bickett and Bonnie Holtby, were near the end of the line. Hipke says that the group remained well bunched considering the circumstances, which means that Bickett and Holtby at a minimum kept up; both were athletes in superior physical condition. It would have been Bickett's job as squad boss, and Holtby's natural inclination, to use their position at the rear to urge the others forward.

The twelve firefighters did not at first hike at maximum speed, Hipke acknowledged. Putnam's calculation that the slow pace contributed to the fatalities was confirmed by other fire investigators. But neither Hipke nor the investigators blame the slow pace on the women. Hipke says that everyone failed to recognize early enough the danger of the situation. The crew hiked more slowly than the strongest would have alone, Hipke said, because that happens when people walk as a group. Putnam adds that carrying packs and tools, a sure sign that they underestimated the threat, slowed everyone down.

The leaders, Thrash and Roth, could have struck out on their own, as smoke jumpers are trained to do and as they saw Hipke do. Thrash had a reputation for enjoying the company of women on the fire line and had been talking to Hagen just before the blowup, leading to speculation that he had remained with the group out of concern for Hagen and Beck. Friends, though, say that Thrash liked to hang around young people regardless of their gender. Thrash had a family of his own: his wife, Holly, daughter Ginny, ten, and son, Nathan, seven.

Roth, at thirty-one more than a decade younger than Thrash, probably followed Thrash's lead, but he had a caring personality and thus his own reason for staying with the group. Friends say that Roth was always doing something for others, repairing a friend's car or fixing a batch of homemade ice cream or wine to give away.

There is much evidence that Thrash and Roth stayed with the others because they felt responsible for them—they were in the lead, they were older, they were smoke jumpers, and initially they were confident that everyone would make it. There is no convincing evidence that Thrash and Roth slowed down, or needed to, for the women.

* * *

A SEPARATE REPORT on the South Canyon fire, directed to workplace safety, was issued February 8, 1995, by the Occupational Safety and Health Administration of the Department of Labor. The OSHA report flatly blamed fire supervisors, on and off Storm King Mountain, for the disaster.

"This was a management failure," said Joseph A. Dear, assistant secretary of labor for OSHA, in a news release accompanying the report. Dear accused fire managers of "plain indifference" to the safety of their employees, but neither he nor the OSHA report named names.

Dear elaborated in a separate letter accompanying the report: "The primary cause leading to the deaths of the 14 was that no one person was responsible for insuring the safety of the firefighters." The violations cited were "symptomatic of the lack of management attention" to the safety of firefighters.

The most significant charges were that the BLM and Forest Service "willfully" failed to communicate effectively the identity of the incident commander, to establish adequate safety zones and escape routes, to make available weather and fire behavior forecasts and to post fire lookouts and reduce hazards while constructing a fire line downhill.

Though the charges were grave, OSHA has no power to impose financial penalties on other federal agencies as it can on private companies. Chief Thomas pledged the Forest Service "to do all in our power to strengthen our management oversight." Dombeck, speaking for the BLM, promised full cooperation with OSHA to correct the faults.

OSHA's focus on management mistakes provided a balance to the South Canyon Fire Investigation. But the OSHA report was brief, barely over three pages long, and its lack of specifics reduced its impact. By comparison the South Canyon Fire Investigation is 39 pages long, plus 176 pages of appended lists, illustrations and documents.

One explanation for the skimpiness of the OSHA report may be that the agency had just concluded a nasty exchange with the Forest Service over a previous OSHA report on a fire fatality. OSHA had cited the Forest Service for thirteen violations, compared to nine violations for the South Canyon fire, in the death of a single firefighter, Frankie Toledo, killed under comparatively straightforward circumstances on April 22, 1993—while conducting a prescribed fire operation in the Santa Fe National Forest in New Mexico. The Forest Service had responded by challenging OSHA's competence to investigate any fire death. OSHA then had cited the Forest Service again, for failing to take corrective action. Officials of both agencies had been called to Capitol Hill to answer for the squabble.

That episode at a minimum made OSHA more sensitive about writing fire reports. On the South Canyon fire OSHA tried to head off a similar counterattack, hiring William Teie, a retired California fire supervisor, to provide them with an expert appraisal of the fire.

The backup documents for the OSHA report were substantial—nearly six thousand pages, including transcripts of dozens of tape-recorded interviews. Those interviews included two sessions each with Pete Blume and Winslow Robertson, the only subjects to be formally reinterviewed. The first interviews with them are deferential in tone, the later ones downright accusatory.

Blume told OSHA the first time around, on October 6, 1994, that air tankers could have been dispatched to Storm King on July 6 if only someone had asked.

"You know, if they had wanted them, we could have made them available. We could get them from the outside," Blume said. Those remarks went unchallenged.

Three months later, on January 3, 1995, Blume acknowledged that the situation had been far more complex than he had previously acknowledged. "I knew that they did not have as much [personnel and aircraft] as they needed for the fire that they had, but we were trying," Blume said. "It would have been ideal if we could have committed one or more helicopters to the fire prior to the sixth."

OSHA QUESTIONER: And you knew that [Butch] Blanco was asking for more people?

BLUME: Right. There was no question that they wanted more and that we wanted to get them more. . . . If we'd had the people on the fire on the fifth that were there on the sixth, it might have been a totally different situation.

In his initial interview on August 4, 1994, Robertson's role was to support Butch Blanco as a distraught Blanco told his story. When Blanco mentioned his quarter-century of service with the Glenwood Springs Fire Department, the OSHA questioner said, "Really? You don't look that old."

"I am," Blanco said. "I'm aging rapidly here."

Blanco never returned to the fire line, though the BLM held his job open for several years. At one point he told friends he wanted to pull on fishing waders, go to a river and never come back. He eventually married his longtime companion, Sharon, and turned his plumbing business into a full-time enterprise.

During this interview Robertson took over the narrative from Blanco at many points, providing a brief account of his own activities in the process. Robertson's statements went unchallenged—"Keep in mind we don't know squat," the OSHA questioner remarked at one point.

When Robertson was interviewed alone, nearly five months later on December 30, 1994, the questioning turned harsh. "What were you responsible for on this fire?" Robertson was asked. "And who was responsible if you weren't?"

At first Robertson claimed he "didn't have any involvement" in the fire between his first visit to Storm King on July 3 and the blowup on July 6. As the interview progressed, his memory improved. In fairness to Robertson, the interview, conducted long after the event, was done by telephone while Robertson was at home—at one point children interrupt the exchange.

Robertson's first revived memory was of receiving a telephone call from Blanco the night of Monday, July 4. "He said he needed a lot more in the way of resources," Robertson said.

Robertson then recalled an incident not easy to forget in which he visited the Western Slope Coordination Center on July 5 and failed, under humiliating circumstances, to upgrade the priority of the South Canyon fire, by then a sizable thirty-five acres.

"I went over to the Western Slope . . . and I requested that we get additional smoke jumpers and air tankers and helicopters assigned to it. . . . I asked them, 'These are what we need for this fire. I'd like a higher priority placed on it.'

"I was told, 'Well, it's not initial attack,' and I argued with them that it was initial attack. . . . I argued with them that it should be a high-priority fire. We didn't have anybody on it yet, and it was going to need air support and smoke jumpers because it was such a long hike."

He remembered that he had been refused by a dispatcher, but could not remember a name, and offered no explanation why he, an assistant fire-management officer, had accepted a no from someone of much lower rank. The episode has an unreal quality, though something like it might have occurred.

The OSHA investigator at one point asked Robertson what responsibilities he had for the fire before it blew up. "Well, I guess I don't know," Robertson replied.

Who, then, was responsible for "monitoring" the fire from July 3 until July 6? the questioner persisted. "On those days, I guess that would be me," Robertson conceded.

OSHA QUESTIONER: So you guys [Robertson and Blume] knew that the fire was doubling [in size] every what, twelve hours?
ROBERTSON: Well, I don't know. I guess I couldn't say if we knew it was doubling every twelve hours. We knew that it was, you know—
QUESTIONER: But you knew that they weren't getting a handle on the fire.
ROBERTSON: Yes.
QUESTIONER: That they didn't have control of the fire.
ROBERTSON: Yes.

After the September rains arrived, Blume, Robertson and others at the BLM's Grand Junction District took a moment to assess their performance during one of the worst fire seasons in memory. It was time, they decided, to start feeling good about themselves again.

They gathered at the district office for announcements and socializing. Someone sent out for pizza. A memo went up on the bulletin board:

"Subject: Superior Performance.

"We wish to recognize and commend the outstanding effort that you and your fellow workers made during the 1994 fire season. With all that has happened this season, it may be difficult to feel good about this year. Nonetheless you should take great pride in the many positive contributions you made during a period of severe fire."

The district had suffered through twice the normal number of fires, five times the average acres burned, plus the "emotional trauma" of the South Canyon disaster. "As time helps to heal the scars left by this season we hope that you will take satisfaction in having done a great job. Thank you!"

Then came announcements. Several managers would be receiving recognition they could take home after the party: pay raises! Not onetime bonuses but salary increases that would follow them into retirement.

Among the favored few were Pete Blume and Winslow Robertson; Blume's annual salary went from $38,129 to $39,285, and Robertson's from $31,512 to $32,466.

EIGHTEEN

Aｆｔｅｒ ｔｈｅ ｂｌｏｗｕｐ had passed, a single blackened pine tree stood like a sentinel at the top of the west-flank line, branches outstretched in a cross. These branches, weakened by fire, broke off over the winter, and by spring the tree had become a spike.

This story cannot end without returning to the west-flank line a last time before the legacy of those who died there vanishes. Lessons remain that only they can teach, accounts that only they can provide. Their fate raises questions that go beyond fighting fire: When do you risk everything for others? When do you go it alone? The preservation of evidence at Storm King, thanks to the memory of Mann Gulch, means that those who paid the highest price will have a say about what happened to them in their final minutes.

By the time Jim Thrash halted on the west-flank line and said "Shelters," the air vibrated as though a locomotive were roaring out of a tunnel. A crescent of fire spilled over the spur ridge behind him and the others. A spot fire appeared up-gulch, multiplying in size at a cancerous rate, doubling and doubling again. The radio blared, "Get the hell up here, *now*!"

What was Thrash, the oldest, most experienced jumper on the

mountain, doing? There was space enough for one or two shelter deployments around him, but no more than that. Did Thrash plan to shelter up himself, leaving the others to fend for themselves?

Thrash very likely had a plan but no way to pass it along in the turmoil—when Wag Dodge yelled at his crew to step into his escape fire in Mann Gulch, they thought he had gone crazy. Thrash pulled out his fire shelter; a few years earlier he had watched a training film that showed firefighters using shelters to protect their backs while running from a fire. He had talked over the film with another McCall smoke jumper, Steve Mello, and said he planned to use a shelter in the same way if he ever had to run for it.

Thrash's partner, Roger Roth, was close enough to hear Thrash say "Shelters," and he alone might have understood what Thrash was doing. Eric Hipke, directly behind those two, could not understand what Thrash meant, and he bolted. Hipke, a match for the mountain, powered up the ridge. Two years later, at the same minute on July 6, Hipke would be at the same spot where he had left Thrash and Roth. This time he would be bent over pruning Gambel oak from the old fire line, unaware of the moment; Hipke had returned to smoke-jumping after his burns healed and was assigned in 1996 to the jumper base at Grand Junction, where in off-duty time he took a hand in keeping the west-flank line from being overgrown.

When Thrash and Roth halted, Don Mackey, at the rear of the line, must have wondered why everyone was bunching up this near the goal, the ridgetop 150 yards away. If the leaders had found a good spot to deploy, it was invisible to him. Maybe the Prineville squad bosses—Jon Kelso and Tami Bickett, a few steps ahead of him—knew what was going on. The march resumed before Mackey could move up.

It was approximately 4:11 P.M., a benchmark in the sequence of events that followed. From this time on, all on the west-flank line could see that they were losing the race with the fire.*

---

* The best official chronology of the South Canyon fire is contained in a report, "Fire Behavior Associated with the 1994 South Canyon Fire on Storm King Mountain, Colorado," issued by the Forest Service in 1999. Most of the work was done at the Intermountain Fire Sciences Laboratory in Missoula, an arm of the Forest Service established in part as a result of the Mann Gulch fire. Following release of the South

On the slope of Hell's Gate Ridge, above the western drainage, Kevin Erickson stuffed his camera down his shirt and began climbing to the ridgetop at about 4:11 P.M. Brad Haugh scrambled upward ahead of him. Farther along the slope Sarah Doehring and Sunny Archuleta finished taking photographs of the west-flank line and turned to leave.

On the ridgetop Butch Blanco's radio call reporting the blowup to the BLM office in Grand Junction was logged at exactly 4:11 P.M. At that moment Michelle Ryerson and her BLM–Forest Service crew, and Bryan Scholz and half the Prineville Hot Shots, were being turned back by the rock outcropping on the ridgetop. Rich Tyler and Rob Browning were at the helispot, H2, preparing to run for it.

On Lunch Spot Ridge Tony Petrilli and seven other smoke jumpers were making their way toward good black, while Dale Longanecker climbed onto the ridge from the Double Draws.

The blowup at this time advanced on the firefighters on the west-flank line from three sides. The main arm of the blowup had crossed the spur ridge and now rolled up the hill behind them. The spot fire farther up the western drainage hooked toward them across open ground. The wave of flame swept toward them from the opposite, Colorado River, direction.

The arms of the blowup began to merge into a gigantic U or horseshoe of flame, with the twelve at its center.

Canyon Fire Investigation, the lab was asked by Ted Putnam to make a further study of the fire's behavior. Putnam and Dick Mangan participated in the project, which took more than three years, far longer than originally planned. Much of the extra time was spent piecing together the chronology.

The others involved were Bret W. Butler, Roberta A. Bartlette, Larry S. Bradshaw, Jack D. Cohen and Patricia L. Andrews. Sue Husari analyzed the fire's behavior for the South Canyon Fire Investigation (the more detailed lab study confirms Husari's initial finding that the development of the blowup on Storm King was "typical" behavior for fires such as this one). The lab chronology traces actions for the fire, the weather and the firefighters from July 2 through July 6. The study deliberately does not go into matters such as why events happened, who was responsible or what lessons should be learned. Instead it seeks to provide a reliable foundation of fact so others can pursue those questions.

In the interest of full disclosure, it should be said that the author shared interview transcripts, notes and observations with these investigators.

As they started the final climb, the fire held the advantage. It ran up the slopes at speeds of from 5.5 to 10 miles an hour and perhaps much faster—the wind might well have slapped the smoke column flat against the ground in some places, creating instant fire.

In reenacting the race between fire and firefighters, Jack Cohen, the fire scientist, hiked at speeds ranging from 2.4 miles an hour on level ground to barely over a mile an hour on the steepest portion, the fifty-five-degree rise near the top of Hell's Gate Ridge. The blowup, then, chased the firefighters at speeds at least two to five times faster than a well-conditioned man could move over the same terrain.

No one who lived saw what happened to the twelve after Hipke took a last glance back and locked eyes with Roth at about 4:11 P.M. and 30 seconds, noting deep concern but not panic in Roth's expression.

Thrash and Roth halted in a cleared space next to The Tree seconds after 4:12 P.M., with advance waves of heat, sparks and smoke swirling around them.

SCOTT BLECHA, THE ex-Marine and Prineville's strongest man, saw a last chance to break away. He must have watched in an agony of frustration as Hipke took off for the ridgetop. Blecha was nearly a physical match for Hipke, but unlike him, Blecha was stuck in the middle of the group. Bulling his way past half a dozen others was no option. The fire line widened at The Tree, and for Blecha it was now or never.

He made his move, leaving a gap behind him. He charged up the slope carrying his heavy, homemade root ripper, legs driving, arms pumping, a bigger-than-life role model. No one followed him. The flat top of Hell's Gate Ridge came into view in slow motion, bit by bit.

He was halfway there when a pulse of superheated air struck him, rolled on and flattened Hipke several dozen yards ahead. The root ripper fell from Blecha's hands. He staggered forward another seven yards until, forty yards short of the ridgetop, the fire drove him to his knees. He remained upright on his knees, facing the top of the ridge, gradually bending down until his forehead touched ground.

The time was approximately 4:13 P.M. and 45 seconds.

THRASH STRUGGLED TO deploy his shelter in the howling wind at The Tree. Others around him were having trouble getting their shelters out, but there was nothing more he could do for them. Thrash had stayed with the group, had tried to tell them to ready their shelters. But everyone was spread out. The noise was awful.

Thrash had his shelter over his back when a blast of heat struck him. The heat partially deflected off the shelter; it melted the straps of his backpack, which slipped toward his feet. Fusees carried in an outside pocket of the pack caught fire, igniting at 375 degrees but burning at 1,700 degrees, well above the melting point of shelters. Afterward, fusee slag was found along Thrash's left side and between his feet.

The blast staggered Thrash, but he managed to deploy his shelter properly, his feet to the fire, the shelter covering his body. The hissing fusees went with him into the shelter. The hold-down straps at the foot of the shelter burned through, allowing the shelter to come loose. The shelter began to turn inside out; it rolled up, exposed Thrash's body and piled up around his head.

Thrash's deployment became a vital lesson of the South Canyon fire. Firefighters previously had been taught to take their packs with them into fire shelters. Thrash's deployment reversed that message. Firefighters now are taught to leave packs outside shelters, taking just a water bottle inside. This not only avoids a problem with fusees but makes it easier to get under a shelter. That lesson, which may well save lives in the future, would not have been learned at Storm King if Thrash's body had been removed from the mountain the night of the blowup and the evidence of his deployment destroyed.

The two rookies, Kathi Beck and Terri Hagen, standing near Thrash, never fully deployed their shelters. Hagen's shelter was torn from her hands and found partially opened four feet above her body. Beck managed to take her shelter out of her pack but was not able to open it. She was found next to Thrash, facing downhill.

A FEW STEPS below Thrash, Roger Roth and Doug Dunbar tried to share the same shelter. They were found side by side, with Dunbar's right hand touching Roth's back. Roth had turned away from Dunbar, and most of the shelter was over him. Dunbar's body lay exposed on the ground, but his clothing was in better shape than that of others, indicating that Dunbar was under a shelter for a time.

The likeliest explanation is that Roth had his shelter out as protection, imitating Thrash, and thus was able to deploy quickly. Roth was under his shelter when the fire struck. Dunbar crawled in with him. Roth then turned away, perhaps recoiling from the heat that followed Dunbar, and pulled the shelter off Dunbar.

Once flames could work on both sides of the shelter, the aluminum siding separated from the fiberglass backing. The aluminum melted when the temperature reached 900 degrees. The heat was intense enough to melt the knobs on Roth's radio under his body but did not silence it. The radio remained "on and fluttering," Putnam later observed, capable of receiving the message Wayne Williams heard after the blowup, calling his name.

The practice of sharing a fire shelter is a familiar one to fire investigators, who credit it with having saved lives in the past. "If this sharing compromised the safety of [Roth], this is the first known incident," Putnam said. Doug Dunbar's mother, Sandy, whose nightmare in which a firefighter had stared at her with a blackened face now seemed to be foresight, believes that it could have been the other way around: Doug Dunbar could have deployed, and Roth could have tried to share Dunbar's shelter. A second, unopened shelter was found at Dunbar's feet. Putnam and other fire investigators believe Sandy Dunbar's explanation to be the less likely one.

DON MACKEY MUST HAVE become desperate to talk to the leaders, Thrash and Roth, or to the Prineville squad bosses, Bickett and Kelso. Too much was happening on his fire without him. But he was stuck at the rear of the line.

Mackey was thinking, Go, Go, but as the group began the final ascent, it slowed down. Mackey had just started the climb when everyone stopped. Was this it? Were they deploying? Bickett and Kelso, once pledged to marry and now best of friends, were together ahead of Mackey. He joined them in a few strides.

The blowup, a swirling mass of flame and primeval darkness, struck with the force of a comet. It slammed Bickett and Kelso to the ground. Mackey avoided the full fury of the blast; either he was kneeling to take out his shelter, or Kelso's and Bickett's bodies acted as shields. Something protected Mackey, because a great deal of evidence indicates that after the initial blast, he stood up.

He had never been more alone. Smoke and flames engulfed him in dizzying waves. The truest fear of death, the knowledge that death is imminent and unavoidable, pressed on him from every side. Such fear sends a torrent of chemicals raging through the body, numbing every thought except concern for self. Considerable evidence indicates that Mackey did not give in to this fear. For that to be true, though, he would have had to concentrate on something other than himself.

He left his pack behind—it was found later near Bickett and Kelso—and turned downhill, the opposite direction he would have taken for his own survival. Perhaps he was staggering blindly, but he made it a long way for someone in that condition. He passed Levi Brinkley, knocked flat before he could open his shelter. A few steps farther along he came upon Rob Johnson, Tony's brother, who lay on his stomach, holding his chain saw up in one hand.

At this point flames melted a nylon knife case on Mackey's belt, and the knife handle and three blades fell to the ground. The handle and two of the blades were found by the MTDC team during their examination of the site. The third blade was found a few days later by Mike Mottice, the BLM area-resource manager—Mottice did not mark the spot but said it was roughly the same place the other parts were found.

Considering the number of knife pieces—four—it is unlikely

they slid or were kicked to the same spot from somewhere else. The knife was special, a replacement for one Bob Mackey had given his son when Don had become a smoke jumper. Don had lost the original knife after carrying it on 154 of his career 155 parachute jumps, and Bob had replaced it with a twin knife he had bought for himself. There is no question it was Don's knife, and he was seen wearing it on his belt that day. The location of the knife handle and blades, then, marks Mackey's path.

He had made a long journey to arrive at this place: thirty-four years of life; a marriage and two children, one of whom, Leslianne, was to be six years old the next day; and eight years in the smoke jumpers. He had chosen to return along the west-flank line, and perhaps he chose to turn back this last time to check on others; it would have been like him.

He found Bonnie Holtby, the last in line, twenty-five feet below the spot he had left his pack. Holtby was curled up on the ground beyond help. The duty left to Mackey was to protect himself. He dropped to the ground and put his head on folded arms in the classic defensive posture, the only one of the fourteen found in that position. Or, alternatively, the extreme heat might have caused his arms to contract; there is no way to settle the matter. He and Holtby faced away from each other, their legs close enough to touch.

The time was approximately 4:13 P.M. and 30 seconds.

There are many ways to die from fire, but intense, dry heat may be the quickest. Dr. Marella L. Hanumadass, known as "Dr. Dass," has seen the full spectrum of fire deaths as director of the burns center of Cook County Hospital in Chicago, one of the nation's busiest burn clinics.

When a blast of dry heat enters the air passages, Dr. Dass said in an interview, it sears the delicate tissues in the throat above the Adam's apple. The insult causes the muscles that control the breathing reflex to contract like a fist, eliminating the need to draw breath and cutting off the supply of oxygen to the brain. When this happened to Mackey, he became unconscious in seconds and died moments later.

Afterward, many families comforted themselves with the thought that their loved ones at least died quickly. The autopsy reports on the fourteen list the same cause of death for each, "asphyxia," or suffocation as a result of inhaling fire, confirming that the end was swift.

NINETEEN

THE FIRST PUBLIC eulogy for the Prineville Hot Shots occurred on a hot, blustery Saturday, July 9, while the South Canyon fire continued to burn. Nine horses with riderless saddles, for the nine hotshots, were led through the streets of Prineville as part of the annual Crooked River Roundup Parade.

After the parade a memorial service was held on the Crook County High School football field. About fifteen hundred people filled the stands; family members stood along the track. In the background a ribbon of smoke rose from a fire being fought far to the southeast in the Maury Mountains.

Oregon's governor, Barbara Roberts, was a few words into her remarks when she had to stop to collect herself. "I remember my husband, Frank, telling me as he was dying, there are only . . ." For nearly half a minute there was silence except for the pop of wind slapping the microphone. Roberts continued, "There are only two assets we have, love and time. Lives are measured by how we spend our time and how we spend our love. These nine gave the two things they had most: time and love."

The bodies of the nine Prineville Hot Shots were flown home

four days later in a vintage DC-3, the workhorse of the skies, the same model airplane used to fly the smoke jumpers into Mann Gulch. In the following days those who died on Storm King were remembered in hometown eulogies.

Kathi Beck

The girl with the flaming, autumn-colored hair should be remembered, said her mother, Susan, for her "courage and great respect for Mother Earth."

Tami Bickett

The beauty pageant princess–turned–firefighter was memorialized in a display at St. Edward Catholic Church in Lebanon, Oregon: a rhinestone tiara, a volleyball and a blue Prineville T-shirt.

Scott Blecha

The iron man of the Prineville Hot Shots was "above all this, a warrior, from the time he was conceived until the time he passed away," said a friend, Marine Corporal Keith Allan Campbell.

Levi Brinkley

"The eldest of triplets loved life in the fast lane, skydiving, rock climbing, bungee jumping and fighting fires," said a friend, Terry Kroger.

Robert Browning

The quiet Southerner was dedicated to caring for the forest and serving people—"he was just a good kid," said his stepfather, Donald Lee Radford.

Doug Dunbar

The honor student took time to help his sister, Rebecca, with her math homework: "I thought he was a genius because he could multiply—my brother was my hero."

Terri Hagen

The girl who always showed up for science lab with a "strange and exotic" insect to study lived life to the full every day, said the lab manager, a friend, Steven O'Connell.

Bonnie Holtby

The caring girl who had decided in 1991 to commit her life "to the Lord," was said by her track coach, Jim Erickson, to be a person of "really strong character and integrity."

Rob Johnson

The brother of Tony, who had escaped the fire by heading down the East Canyon, was insightful about others but forgetful about himself: "He could never keep track of his wallet, but being a CPA, he always knew exactly how much he'd lost," said the Reverend Bruce Russell.

Jon Kelso

The smoke jumper–turned–hotshot, everybody's younger brother, was remembered for his attitude: "Don't hurry, don't worry; take time to smell the flowers."

Roger Roth

The one always looking for things to do for others once told the Reverend Craig Anderson, "When you're standing in the doorway of the airplane preparing to jump, you're closer to God."

James Thrash

The oldest smoke jumper was a "shepherd" who protected others, said a friend, Joe Fox, one of the jumpers who went to Hell's Gate Ridge after the blowup the night of July 6.

Richard Tyler

The rising star of helicopter firefighting was a "model employee, dedicated, devoted—he would go the extra mile for nothing, for no pay," said his boss, Paul Hefner.

GREEN, WELL-IRRIGATED fields mark the serpentine course of the Bitterroot River from its headwaters near Darby to its junction with the Clark Fork River near Missoula. Above the fields, tan meadows and timbered hills rise steeply toward a jagged wall of mountains, the Bitterroots, home to mountain goats. The mountains are broken at intervals by canyons, the most dramatic being Blodgett Canyon, which leads into the heart of the mountains.

Hamilton, population 4,200, an old-time logging and ranching town, lies halfway down the valley. There aren't many streets, but they are broad; the homes are modest but tidy. The Dowling Funeral Home is located in one of the oldest and largest houses. The memorial service for Don Mackey, a favorite son of the valley, drew an overflow crowd of more than 750 to Dowling's. Folding metal chairs had to be set up outside on the lawn, shaded by stately trees. Inside, the chapel was dimly lighted. Although there were tears, Mackey had touched people with laughter and encouraging words. A succession of those who knew and cared for him came to the podium, each with a story to tell.

Don Mackey had lived out the Western dream, roaming, hunting and fighting fires in the mountains. But he was remembered most for his ability to connect with people—rich and poor, simple and complex, children as well as adults. He liked babies, "the smaller the better." He loved his family, but had a "special bond, a friendship that ran deep" with his father, Bob.

Don had a way with animals, especially horses. His horse, Chester, "thought he was the family cat."

He knew himself so well that "he didn't have to try to be who he was." He helped his friends "live a better life," added "quality" to their daily existence and held them to a high standard of behavior and workmanship.

For Don "doing the right thing was always so simple he never had to think twice about it." On Storm King Mountain "Don was where he wanted to be; he chose to be there." He was the "heart and soul of the smoke jumpers."

Life had been fuller and more fun with Don Mackey along.

On the happiest hunting trip he ever took, said Tony Petrilli, Mackey's sometime roommate and fellow smoke jumper, he never saw an elk. After a day or two of fruitless hunting, Petrilli called on "the expert," Mackey.

"We got in the old Power Wagon and headed down the road to some secret area," Petrilli told the assembly. "On every ridge, coming around, Don would say, 'Get ready, get ready.' And of course every ridge we came around, no elk. We'd come through a gully, and Don would impersonate poor George Bush—Oh *please, please* let there be an elk.

"I've come back from hunting trips where my legs ached, my shoulders ached, just dead-beat tired. But I've never come back from a hunting trip where my guts hurt from laughing so much."

Tony and Don's relationship had its more serious side.

"Don and I also shared some tough times together. One thing he told me was, Just be yourself. On July 6 Don was being himself. He didn't have to come back to my group of smoke jumpers to make sure we made it to a safe spot. But he did. He didn't have to go back down the line to try to get everyone else out, but he did. He was just being himself."

Mackey's legacy is alive on the fire line, said his girlfriend, Melissa Wegner. "There's a family in firefighting," said Wegner, who spoke at the memorial service and afterward. "And one of the bonds is Don. And one of the bonds is those of us who know Don. I'm deliberate when I say that: There are some people that knew Don and some that *know* Don."

What happened to Mackey and the others on Storm King could have happened to any firefighter, Wegner said, and on one occasion it had almost happened to her. She was with the Flathead Hot Shots fighting the Dude fire in Arizona in 1990 when the ground at their feet seemed to burst into flame as pine needles ignited. Everyone began yelling to drop tools and packs and run for it. Two Flathead Hot Shots, John Herron and Chris Schustrom, began shouting to the others, "Into the black! into the black!"

"They were way ahead of us, and they stopped," Wegner recounted. "And they waited for all of us. I remember when I passed John, seeing the wall of flame behind us. A lot of people were crying and screaming. A calmness came over me: If I was going to die, it's okay.

"The winds changed, and we didn't die. We sat in the safety zone for I don't know how many hours."

As they hiked out, they passed a burned-out cabin that became for Wegner a symbol of the values—life and property—firefighters strive hardest to protect. "The ash was like velvet on a couch. There was a child's swing set. The seat of the swing was gone. There was a breath of breeze blowing through there. The two chains were swinging independently of each other."

A call came over the radio: Six people who had been passed by Wegner's crew moments before the fire erupted had been trapped and killed.

Wegner's own crew had been lucky, but Herron and Schustrom had helped, giving the others encouragement and direction. If everyone had stopped to help, though, it would have caused a traffic jam with perhaps disastrous results.

"Herron and Schustrom waited and they cared, and I know Don did the exact same thing," Wegner said.

The legacy of Don Mackey, then, has a double edge. When a situation becomes desperate, it can save lives if someone takes a stand and helps others, and such people rightly are admired as heroes. But that role belongs to a few natural leaders willing to risk their lives. Faced with a desperate situation, the best solution for most people, the one that saves the most lives, is to get themselves out of trouble.

Sometimes, though, it's harder to save yourself than to save others.

Don Mackey's ashes were scattered in a grove of ponderosa

pines on the site at the Mackey place where he intended one day to build his own house. A few of his beloved traps hang from the trees. Behind the pine grove rises Blodgett Canyon, and beyond that the Bitterroot Mountains. Above them is the light that makes the mountains shine.

THE BITTERNESS OF losing a child remained long after the memorial services concluded. Some family members began a tradition of visiting Storm King on the fire's anniversary. In 1996 several spent the day on the mountain and then gathered at one of the homes opened to them by townspeople. After dinner on an outside porch, it began to grow dark, and the conversation fell to a murmur. Then one voice cried out, "It's as though they've opened an artery and drained our life's blood!"

The number of returning family members declined over the years. When the Mackeys and others banded together in 1996 to submit wrongful-death claims against the Forest Service, the families of Rich Tyler, Jim Thrash, Bonnie Holtby, Scott Blecha and Jon Kelso did not participate. Such a claim, it was thought, would violate the commitment of their loved ones to fighting fire.

The Forest Service denied the claims, totaling $23 million, in December 1997, on the grounds that full compensation already had been made.

An effort by the Alpine Bank of Glenwood Springs to raise $1,000 for each family to help defray funeral expenses turned into a success beyond everyone's dreams. The bank wound up collecting and distributing to the families more than $432,000.

Glenwood Springs formed the Storm King 14 Committee to come up with a suitable memorial for the town. The committee adopted the motto "We Will Never Forget," after Ken Brinkley, Levi's father, remarked to the chairman, Veto "Sonny" LaSalle, supervisor of the White River National Forest, "Don't let them be forgotten." The committee raised $165,000, plus at least that amount in in-kind donations, and erected a statue and individual memorials for the fourteen at Two Rivers Park. A statue to commemorate all wildland firefighters and individual memorials for the fourteen were erected in the Ochoco Creek Park in Prineville.

Sympathy for the families came at too high a price in one case.

Chris Cuoco, the weatherman, and his family were driving home
after mass on a rainy Sunday, August 25, 1996, when an auto in
the opposite lane skidded on wet pavement, swerved and struck
the Cuoco auto head-on. Cuoco's infant son, Benedict, named
for the saint, was killed. "Now I know how all those parents
feel," Cuoco said afterward. "It's something you can't feel until
it happens."

Less than a year and a half later, on December 30, 1997,
Cuoco's wife, Michele, gave birth to another son, Dominic, also
named for a saint. A note on the birth announcement read, "With
God, all things are possible."

Mann Gulch and Storm King Mountain once were lonesome
places but today draw tourists by the hundreds. Markers explain
what happened in 1949 and 1994. Crosses stand where firefighters
fell. Well-worn footpaths in both places connect the crosses.

In Mann Gulch the crosses, made of concrete reinforced with
steel bars, had deteriorated so badly by the 1990s that Wayne
Williams and the Missoula smoke jumpers sought to replace them
with granite obelisks. When objections were raised to hauling the
old crosses away, it was decided to leave them and simply add
the obelisks, except in the case of David Navon, who was Jewish,
a fact not taken into consideration in 1950 when the crosses were
put up. The smoke jumpers installed the obelisks in the spring of
1997.

Bob Mackey helped organize an effort to erect granite crosses
on Storm King; Terri Hagen's was the exception—her site has
two markers to reflect her Native American heritage, a granite
marker with the crosspiece halfway down to signify the four com-
pass directions and a circle of black steel around a cross to rep-
resent the circle of life.

There was snow on the ground when Bob and Nadine drove
to Storm King from Hamilton the first spring after the fire, in
April 1995, to finish the project. "This was something I had to
do—part of those people is still up there," Bob said. Many volunteers turned out to help; Butch Blanco directed a helicopter
airlift that transported ten thousand pounds of concrete and other
gear to Hell's Gate Ridge.

Nadine had not been sure how she would react to meeting

Blanco. "He was the one who sent Don down the fire line," she said. "I have a rage I don't think I'll ever get over." But when she met Blanco for the first time, at the BLM office in West Glenwood, she hugged him.

Bob and Nadine camped out on Hell's Gate Ridge while Bob, his brother, Tom, and a steady stream of visitors dug holes, poured concrete and placed crosses. Bob was happy to share the work and to stop and trade stories with visitors. But he alone put in Don's cross.

Storm King became a place of pilgrimage. From the first days, when there were wooden crosses at the site, family and friends left tokens—Red Wolf beer cans, snuff cans, a Pulaski, a set of sawyer's earplugs—gestures similar to what happens at The Wall at the Vietnam Memorial in the nation's capital. A Rebel flag marks Rob Browning's cross; a pair of skis are stuck in the ground next to Levi Brinkley's; someone put a Missoula smoke jumper's cap with the handwritten words "Blessed are these who jump fires from DC-3s" at Don Mackey's.

On the first anniversary of the fire the Mackeys and thousands of others—Prineville Hot Shots, BLM firefighters, smoke jumpers, helitacks, personnel from the BLM's Grand Junction District, family members and friends—gathered at Two Rivers Park in Glenwood Springs for memorial activities. Family members addressed the crowd from the same podium where VIPs, including Chief Thomas of the Forest Service and Mike Dombeck of the BLM, were seated.

Marvin Kelso, Jon's father and a schoolteacher in Prineville, praised the character traits of firefighting crews—high self-esteem, love of challenge, a strong work ethic and trust of others. But their greatest attribute, Kelso said, was "having a 'can-do' attitude. 'Can-do' attitudes motivate beautiful people and make beautiful things happen." The audience erupted in applause.

The loudest applause was reserved for a remark by Don Radford, Rob Browning's stepfather. Radford noted that the VIPs had promised to make fire lines safer places. "You people back here," he said, gesturing to the VIPs, "don't talk the talk if you're not going to walk the walk." It brought cries of "Yeah! Yeah!"

Many family members hiked or were shuttled by helicopter to

Storm King. Purple lupines and asters bloomed among the granite crosses. A group of children ran in and out of Bob's legs as he sat near Don's cross, trading stories and an occasional laugh and smile with other families. It would be years, though, before Nadine could say, "I've finally accepted that I'll never see Don walk in my kitchen door again."

The same wind blows over Mann Gulch where young men fell, and Storm King Mountain where the young and not so young, women as well as men, met a nearly identical fate four and a half decades later. The wind that once fanned blowups, and will again, now reaches across the years to join in comradeship those who fell. And they call out to those who follow, Let our sacrifice be enough.

Notes and Acknowledgments

THIS BOOK WOULD not have been possible without the generous help of scores of people. First and foremost I gratefully acknowledge my debt to my father, Norman Maclean, whose book on the Mann Gulch fire, *Young Men and Fire*, opened many doors for this book. I assisted in the posthumous publication of *Young Men and Fire* and became involved in the world of wildland fire as a result. My father is present on every page of this book, or at least his spirit watched over my shoulder and, as was his custom in life, readily offered a word or two of advice.

His spirit, too, was present on Storm King Mountain on July 6, 1994, when the South Canyon fire blew up, mirroring events in Mann Gulch four and a half decades earlier. He would have been enraged over the violation of the unspoken promise of *Young Men and Fire*, that nothing like the Mann Gulch fire should ever occur again. Eventually, I believe, he would have made a peace, acknowledging blowups as part of nature, but with the stipulation that nature includes people willing to confront life's worst catastrophes.

I have tried to make this as true an account as possible of the fire on Storm King Mountain and surrounding events. The dialogue is from people's memories of what they said or heard, or from video and news accounts; the thoughts attributed to people are ones they claimed to have had. Sometimes memories are highly reliable, and sometimes they are not; I have tried to let the reader in on the degree of reliability when this becomes important. But a traumatic event such as a fatal fire blasts away some memories and alters others—two people who took photographs on Storm King during the blowup, for example, lost the memory of having done so, at least for a while.

The events portrayed are drawn from interviews, news accounts, videotapes, photographs, private notes and journals, court records, letters, logs and other documents. I used three collections of materials. Two of these were the supporting documents for separate federal investigations of the South Canyon fire—one a joint effort by the Bureau of Land Management (BLM) and Forest Service that resulted in the first official report, the South Canyon Fire Investigation, and the other an effort by the Occupational Safety and Health Administration (OSHA) of the Department of Labor. Taken together, the documents supporting these investigations come to nearly ten thousand pages. In the text of this book they are identified either by a phrase such as "told fire investigators," "said later" or by a more specific citation where appropriate.

The third and largest collection of documents includes the interviews I conducted with most of the fire's thirty-five survivors and with many other people, and various materials I collected from individuals, newspapers, libraries, government agencies and other sources. My own interviews are identified by a phrase such as "said later" or "said in an interview." More specific citations are given where appropriate.

A few short quotes are included without attribution, though I tried to keep this practice to a minimum. These quotes come from a variety of news reports, both print and video, or from interviews; citing each source, I thought, would have burdened the reader without shedding much light. Where news reports are quoted at length, the source is identified.

I filed more than a dozen Freedom of Information Act (FOIA)

requests with the Forest Service, BLM and OSHA; though the FOIA process can be adversarial, representatives of these agencies handled my requests with courtesy and efficiency. The financial costs of the South Canyon fire reported at the end of Part II, for example, involved a series of FOIA requests over a period of ten months and much digging by agency personnel.

When I started my research, I talked with Chief Jack Ward Thomas of the Forest Service, since retired and now teaching at the University of Montana, who promised his agency's complete cooperation. Such a broad promise inevitably would be tested, and when this one was, Thomas proved as good as his word. I am grateful.

Reporting and writing this book took nearly five years, about fifty thousand miles of auto travel, more than a dozen trips to Storm King Mountain in Colorado and several to Mann Gulch in Montana, as well as trips, sometimes more than one, to Grand Junction and Glenwood Springs in Colorado; Boise, Idaho; Butte, Hamilton and Missoula in Montana; Prineville, Portland and other towns in Oregon; Spokane, Washington; and Washington, D.C. During this time I alternated living at my family cabin at Seeley Lake, Montana, near Missoula, and at my home in Washington, D.C.

I thank everyone who talked to me about the involvement of their family members in the Mann Gulch and South Canyon fires. I offer particular thanks to Bob and Nadine Mackey; Jan Mackey Erickson; Patty Tyler; Lois Jansson, widow of Ranger John Robert Jansson, who was in charge of the Mann Gulch fire; and Patricia Wilson, my neighbor in the Blackfoot Valley and the widow of R. Wagner Dodge, foreman of the Mann Gulch crew.

I thank all who spoke to me about their own experiences, no easy task. Many of these people are named in the text, but several should be mentioned for having undergone repeated questioning: Chris Cuoco of the National Weather Service; Sarah Doehring, Kevin Erickson, Tony Petrilli and Wayne Williams, Missoula smoke jumpers; Bryan Scholz and Bill Baker of the Prineville Hot Shots; Brad Haugh of the BLM; Michael Lowry and Steve Little of the Forest Service; Kathy Voth of the BLM. Bob Sallee, the last living survivor of Mann Gulch, deserves special thanks for once again hiking into Mann Gulch to hash over the 1949 fire there.

Dick Mangan, Ted Putnam and Jim Kautz of the Forest Service's Missoula Technology and Development Center provided unfailing help from the beginning of this project to its end.

A group of smoke jumpers based in Boise, Idaho, and their wives read the manuscript as it developed and offered thoughtful advice and welcome companionship. They are Jim and Karen Kitchen, George and Kathey Steele, Steve Nemore, and Bob and Suzie Hurley. Eric Hipke, the survivor who had the closest brush with death, was part of the Boise group and helped in countless other ways as well, as did Jim Kitchen.

I acknowledge the efforts of Bret W. Butler, Roberta A. Bartlette, Larry S. Bradshaw, Jack D. Cohen and Patricia L. Andrews, Forest Service fire scientists, along with Dick Mangan and Ted Putnam, both mentioned previously, who put together the most accurate official chronology of the South Canyon fire in a study released in 1999, "Fire Behavior Associated with the 1994 South Canyon Fire on Storm King Mountain, Colorado."

The photos of the Storm King 14 that appear on the back endpaper of this book are the ones chosen by the families to be part of the memorials in Glenwood Springs and Prineville. I thank the families and the BLM for making them available.

I also thank Marshall Bloom of Hamilton, Montana; Bob and Billie Jean Burns of Helena, Montana, my cousin and his wife; Mary Challinor of Washington, D.C.; Tim Eldridge, Missoula smoke jumper; David Frey, of *The Glenwood Post*; my editor, Harvey Ginsberg, senior editor for William Morrow and Co. of New York; Jason and Maria Greenlee of the International Association of Wildland Fire; James Kincaid, professor of English at the University of Southern California, Los Angeles; Nick Lyons and his daughter, Jennifer, my agent; Ginny Merriam of the newspaper *The Missoulian*; Robert Mutch of the Forest Service, retired; Larry and Barb Rynearson of West Glenwood, Colorado; Rodgers Wright, a Missoula smoke jumper and one of Don Mackey's best friends; and Garrett Zabel, professor of geology, of Glenwood Springs.

Also Dan Jackson, photographer, of Chicago; Craig Johnson and the staff of the Morton Arboretum in Lisle, Illinois; and Owen Youngman and Rick Kogan of the *Chicago Tribune*.

The unselfish support of my wife, Frances, was essential in more ways than I can count.

Glossary

Air tanker—A fixed-wing aircraft that drops chemical retardant or water on fires.

Blowup—A sudden increase in a fire's intensity and rate of spread that results in a violent, widespread burst of flame.

Bucket drops—Dropping water from a bucket suspended below a helicopter.

Bureau of Land Management (BLM)—The federal agency responsible for low-value, mostly unforested public land in the West and Alaska.

Fire line—A barrier cut or dug along the edge of a fire to deny it fuel.

Fire retardant—A chemical mixture that retards burning.

Fire shelter—An aluminum, tentlike shelter carried by firefighters as protection of last resort.

Forest Service—The federal agency responsible for National Forest and other, mostly forested public lands.

Fusee—A flare used to deliberately ignite fires.

Grand Junction District—The BLM district, headquartered in Grand Junction, Colorado, that includes Storm King Mountain.

Helibase—A main location to park, fuel, load and maintain helicopters.

Helispot—A temporary landing spot for a helicopter.

Helitacks—Firefighters trained to fight fire using helicopters.

Hotshots—Fire crews of twenty who can be flown or bused where needed to fight fires.

Incident commander (IC)—The on-site manager of a fire.

Initial attack—The first suppression work on a fire.

Lead Plane—An aircraft used to scout a fire and to lead air tankers to their targets.

National Interagency Fire Center (NIFC)—The federal center in Boise, Idaho, that coordinates the use of firefighting equipment, crews and aircraft on a national level.

Red Flag Warning—A warning of weather conditions likely to cause extreme fire behavior.

Red Flag Watch—A notice of weather conditions potentially dangerous for firefighters.

Resource Order—An order placed for personnel and equipment to fight fires.

Rocky Mountain Coordination Center—NIFC's regional center, located at Jefferson County Airport outside Denver, for all or portions of five states, including Colorado.

Run (of a fire)—A fast, intense advance by a fire.

Safety zone—An area designated for escape from a fire.

Size-up—The initial evaluation of a fire.

Smoke jumper—A firefighter who parachutes onto fires.

Spot fire—A fire outside the perimeter of a main fire.

Spotter—The person responsible for selecting a jump site for smoke jumpers.

Western Slope Coordination Center—A federal office in Grand Junction, Colorado, that at the time of the South Canyon fire coordinated firefighting resources, especially airplanes and helicopters, for western Colorado and a slice of eastern Utah.

Wildland fire—Any fire outside an urban area or on any open land.

Kathi Walsleben Beck

Scott Alan Blecha

Douglas Michael Dunbar

Levi Brinkley

We will

Tamera Jean Bickett

Robert E. Browning Jr.

Terri Ann Hagen

Bonnie Jean Holtby

Jon R. Kelso

Jim Thrash

ver forget

Don Mackey

Roger Roth

Richard Kent Tyler

Rob Johnson